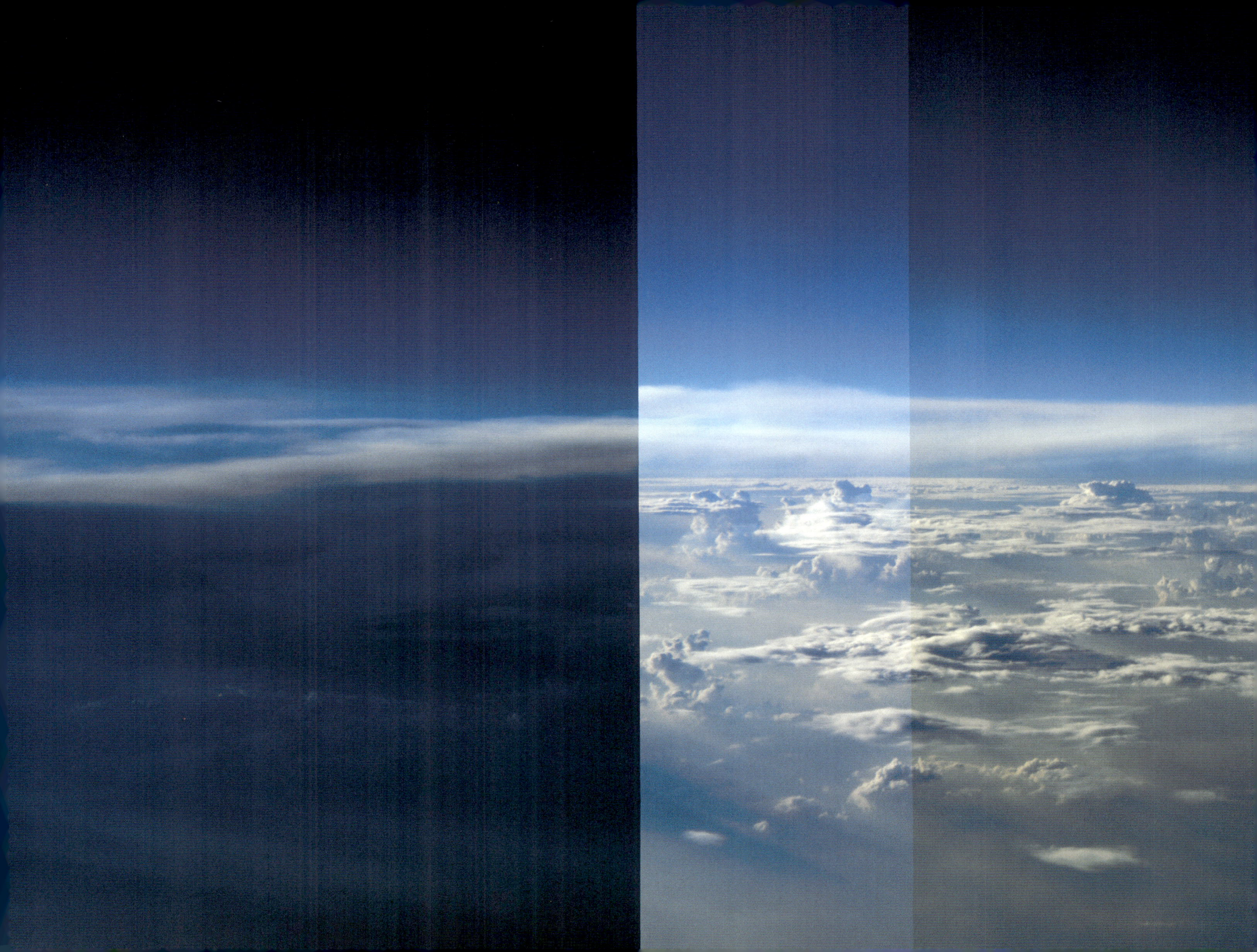

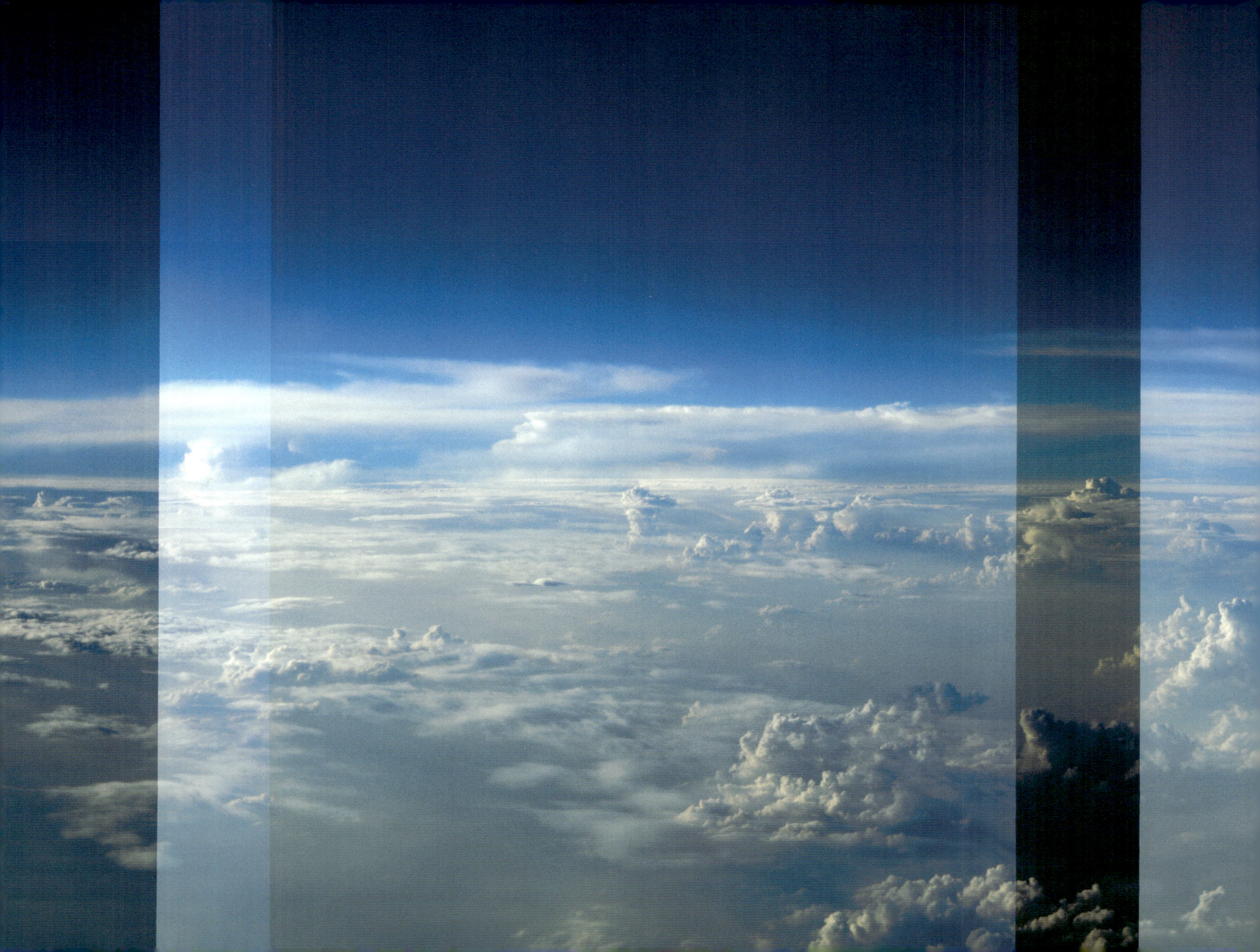

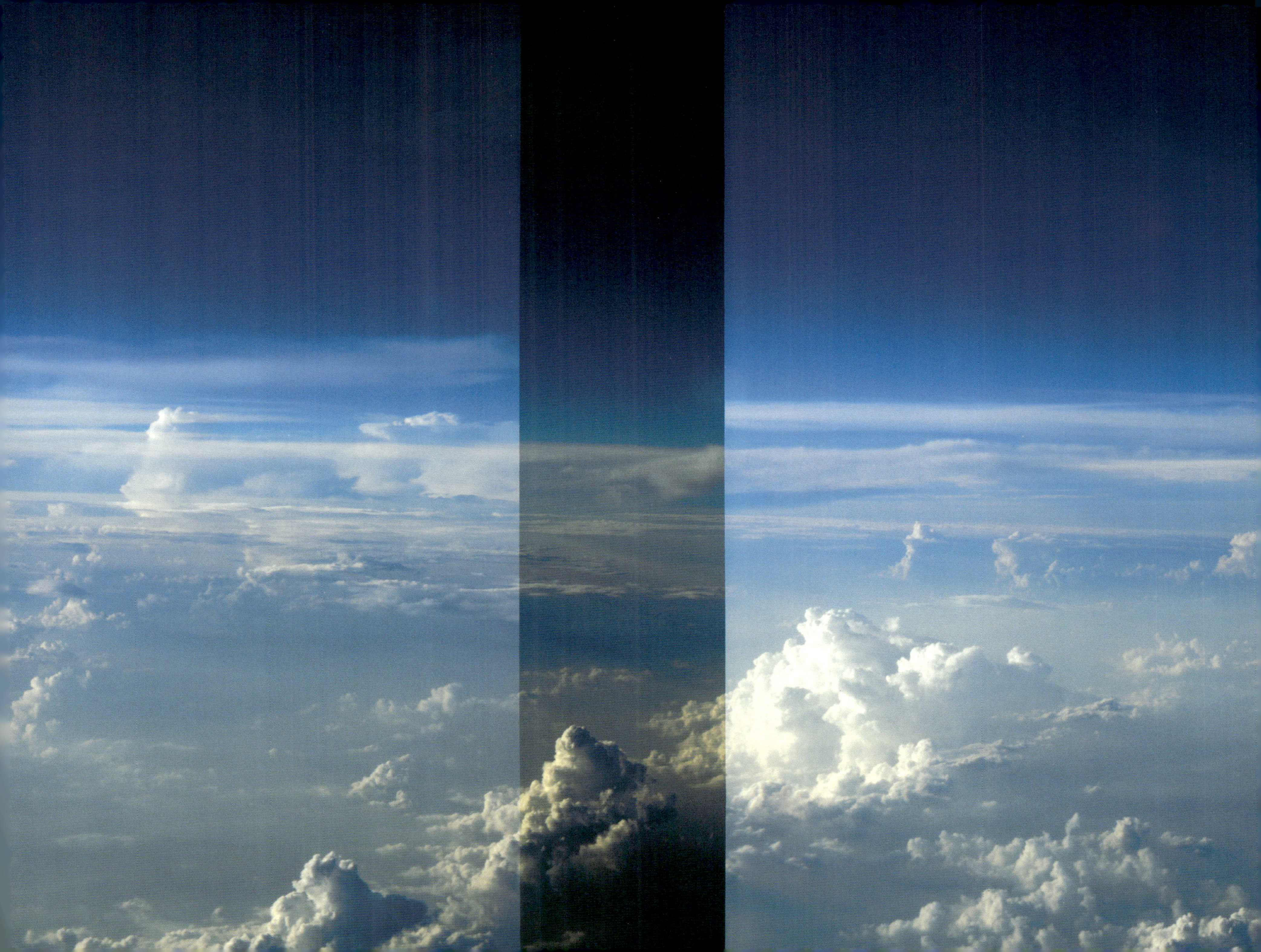

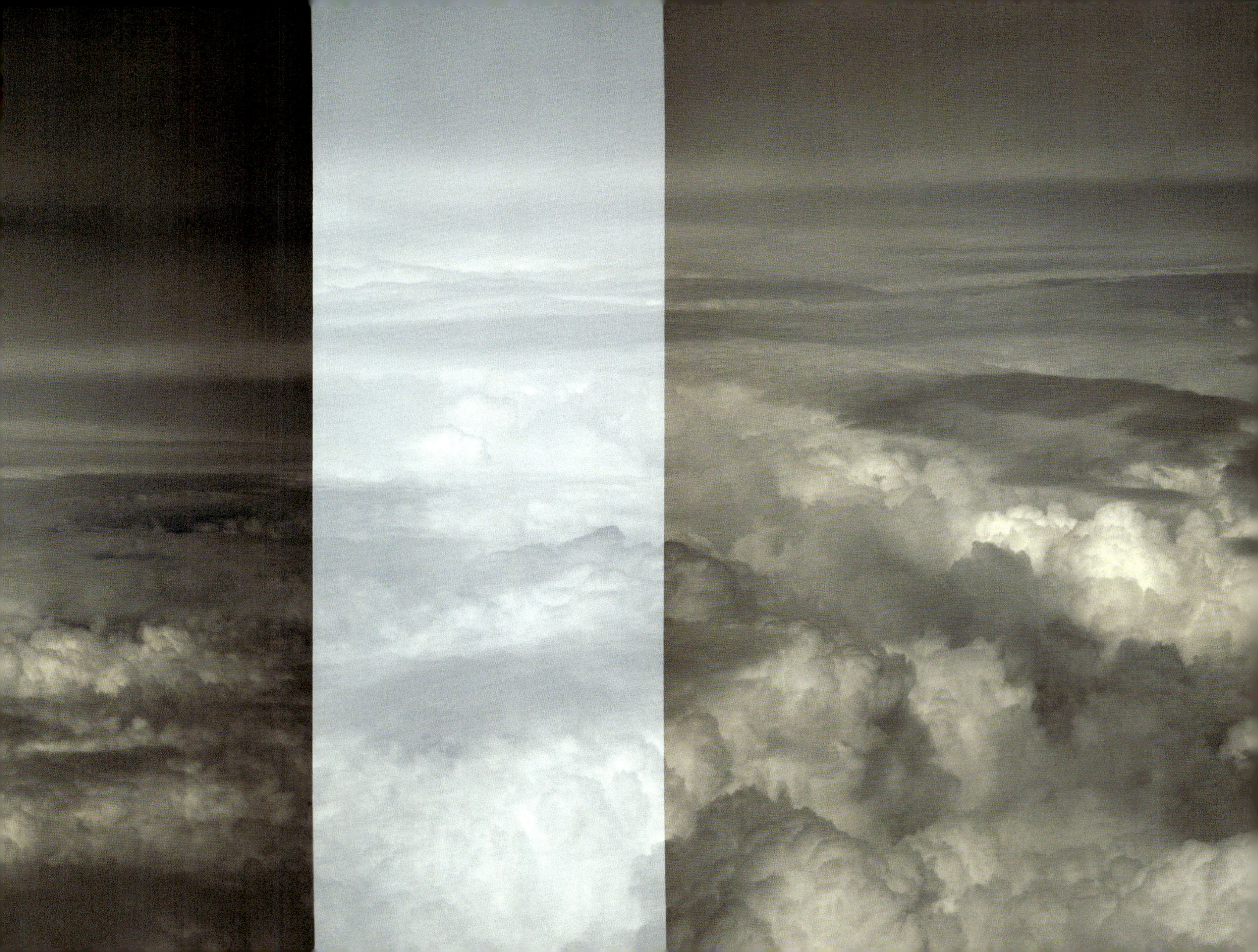

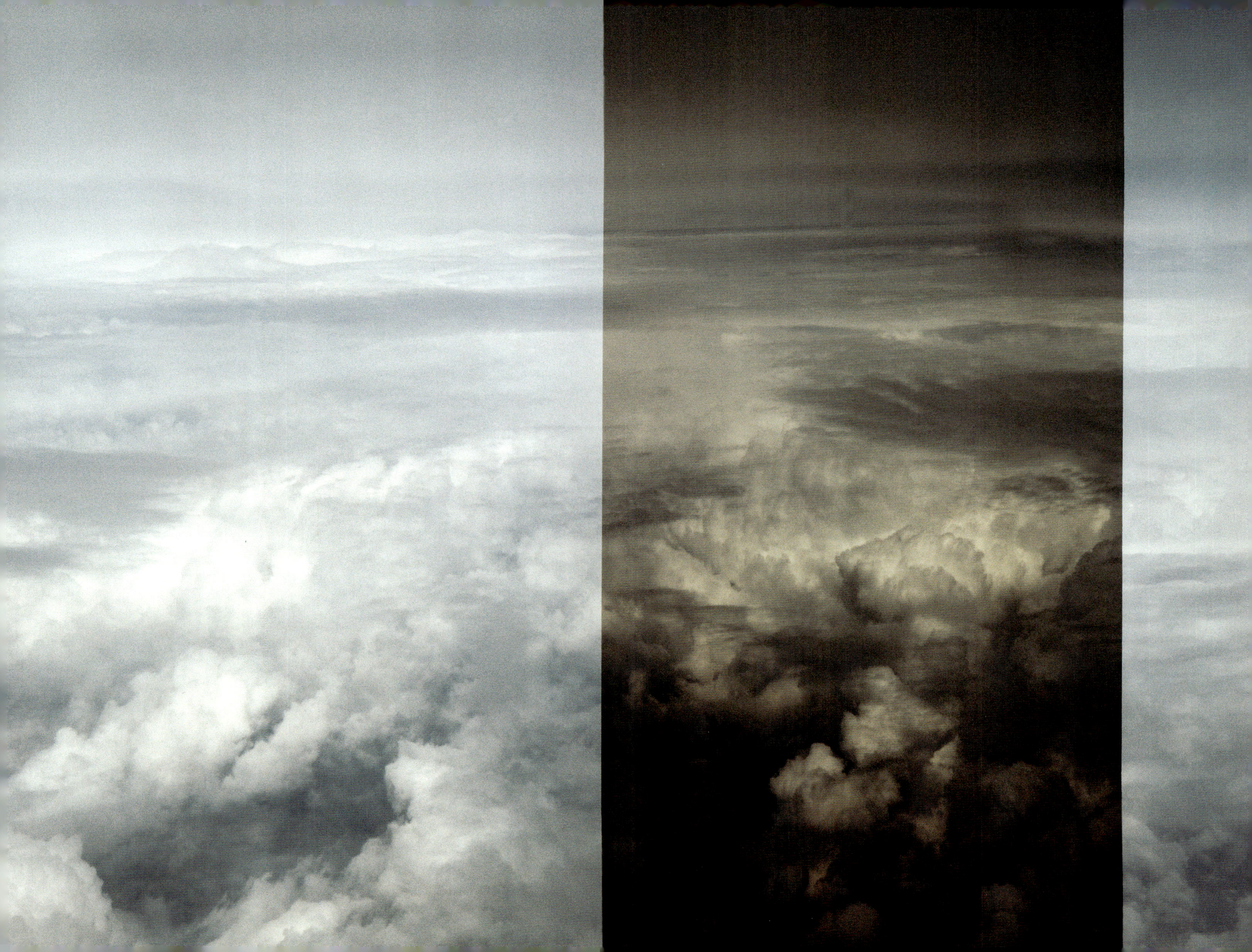

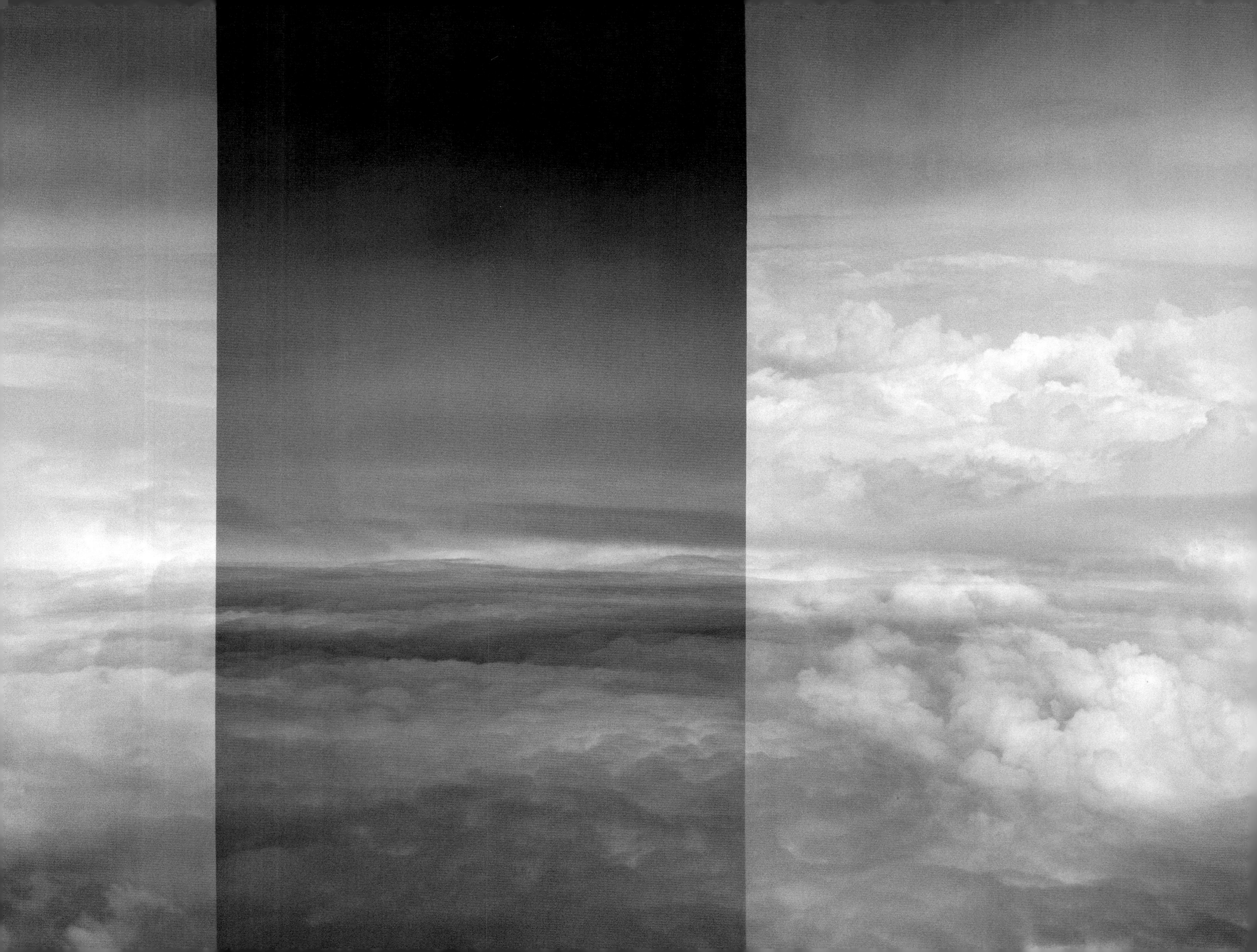

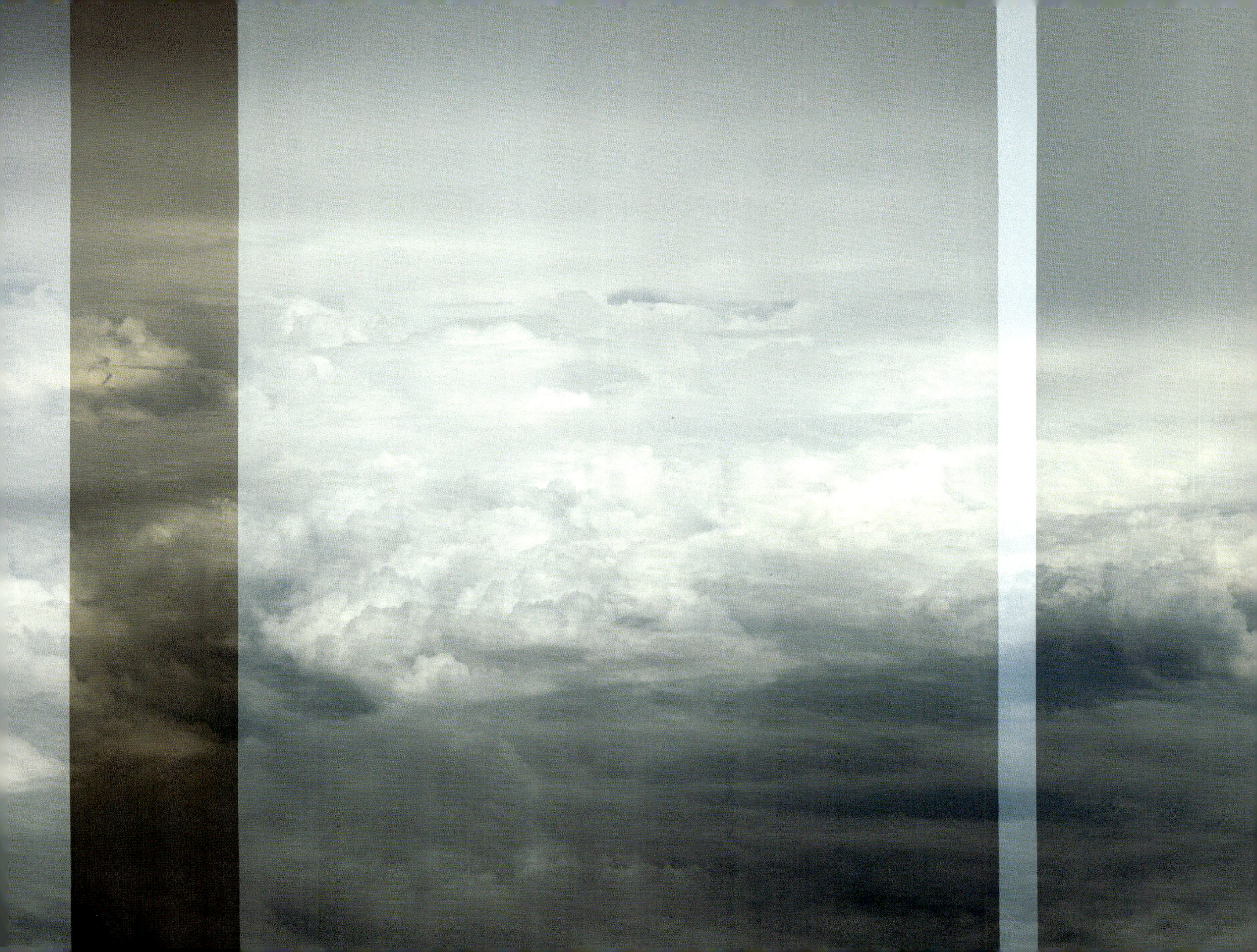

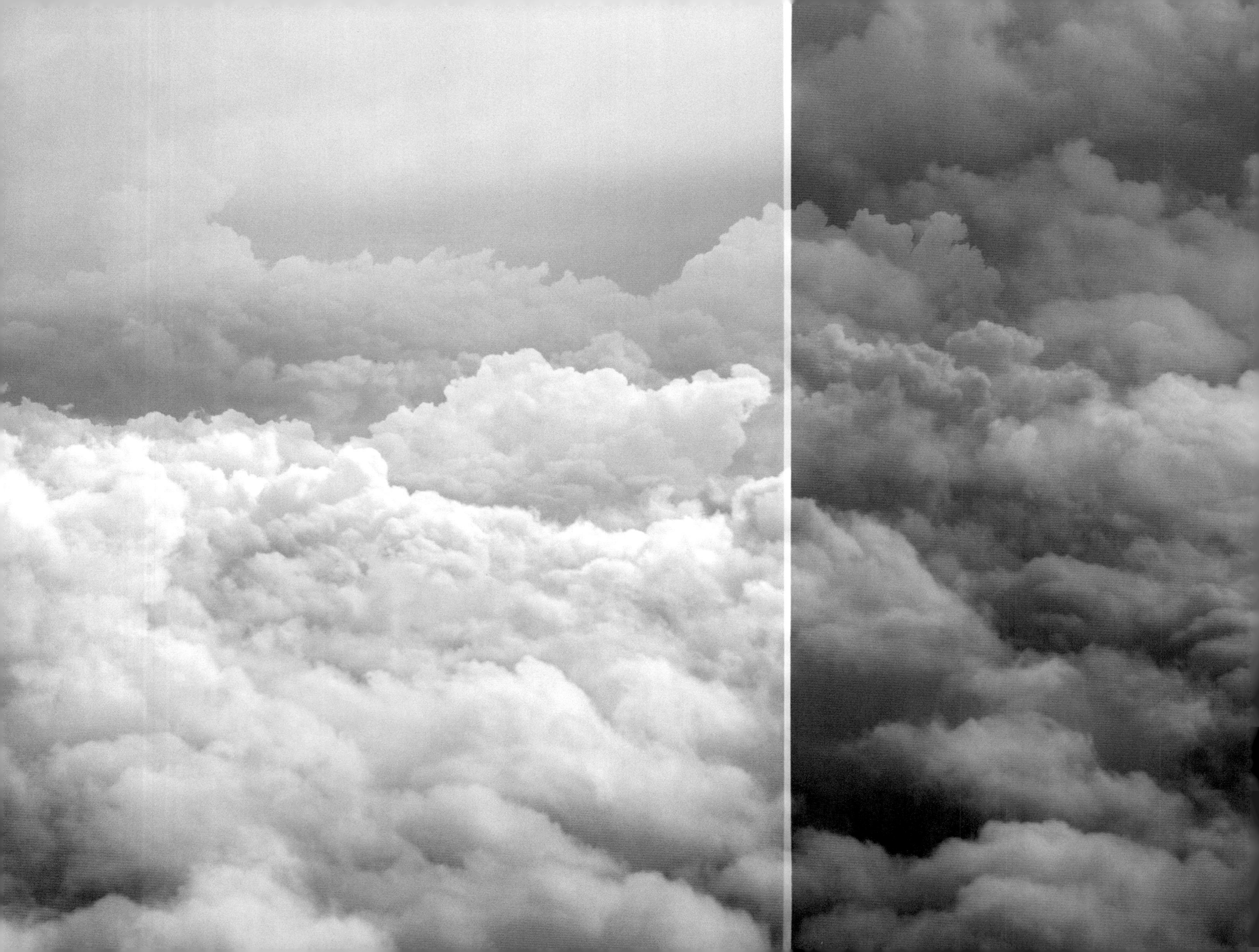

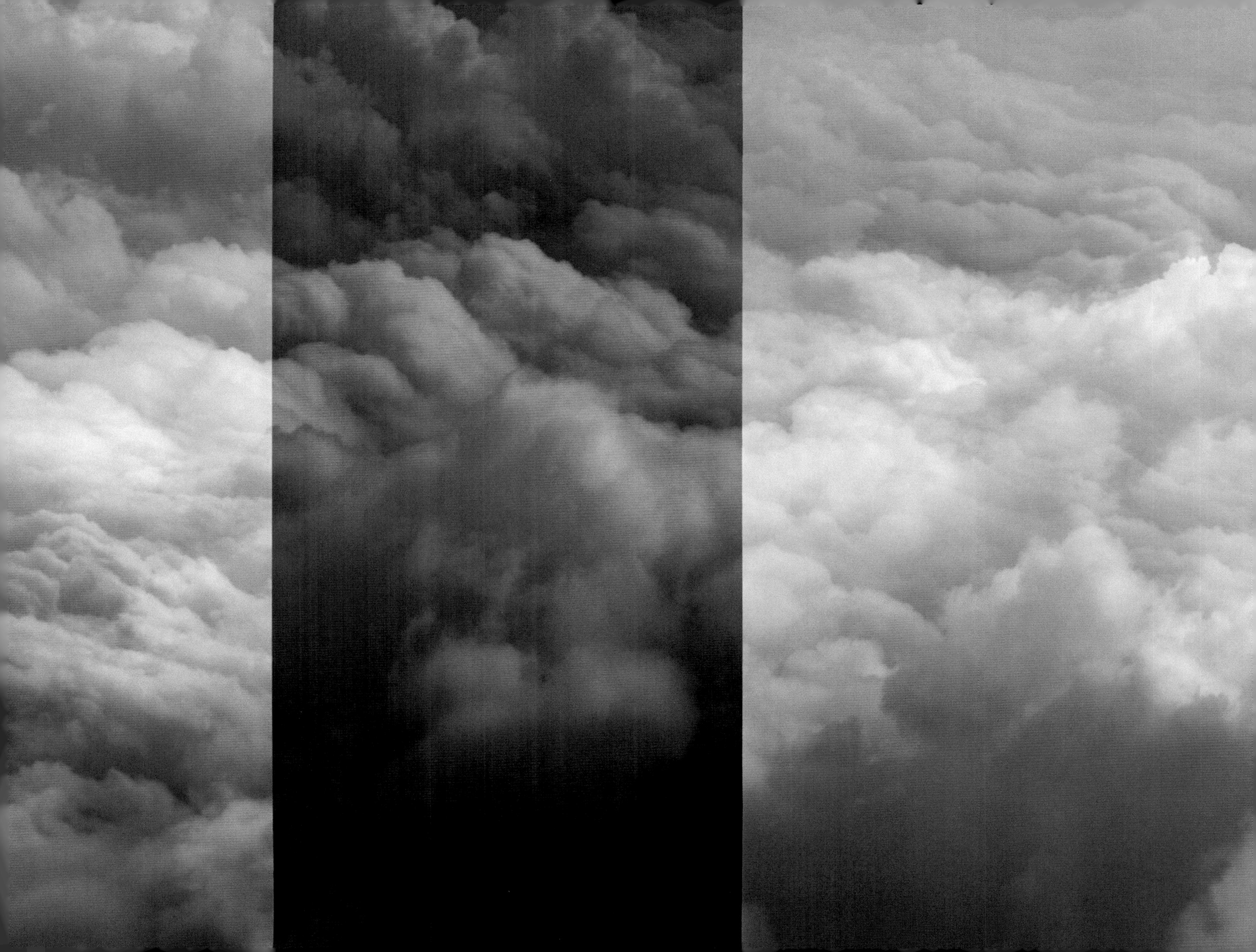

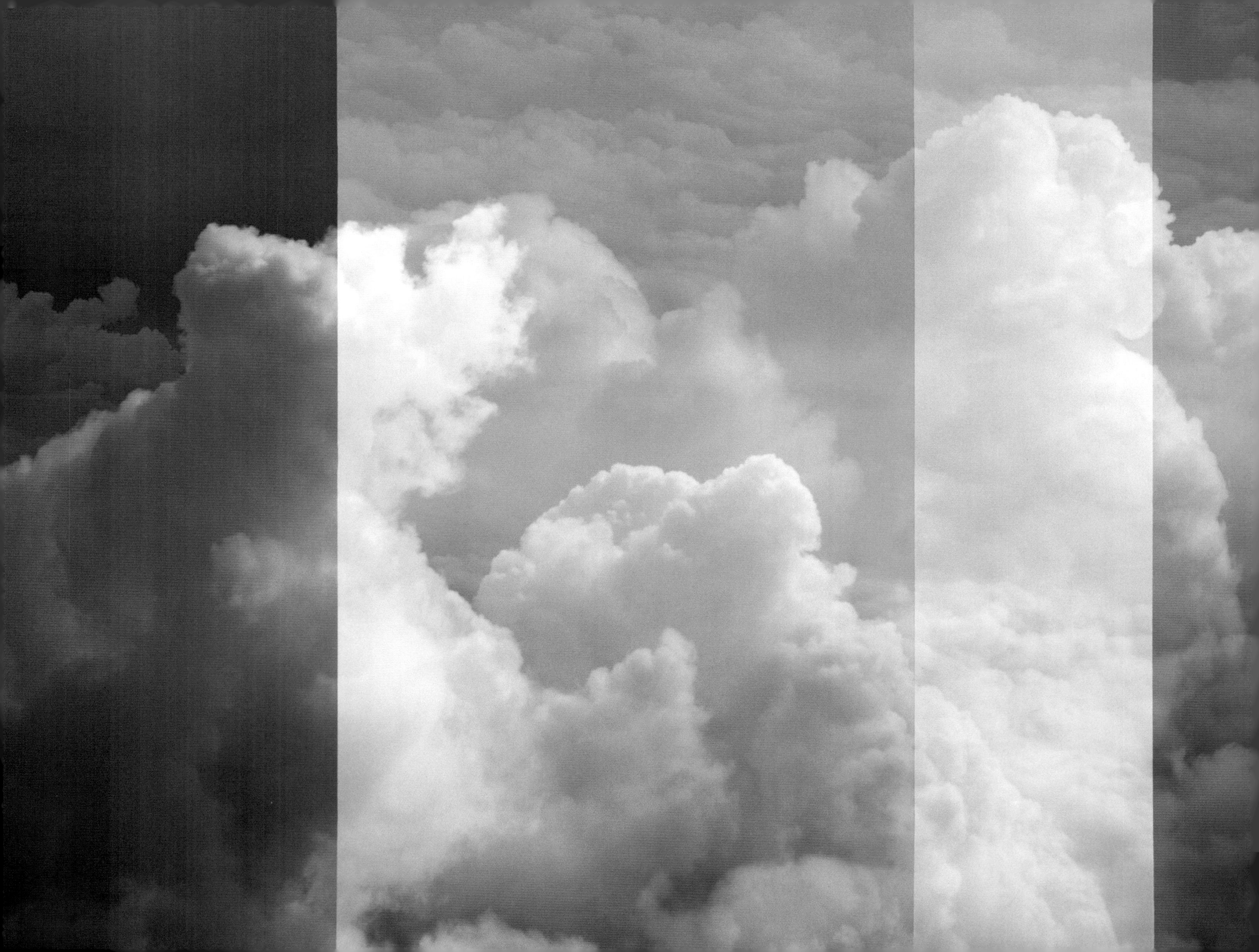

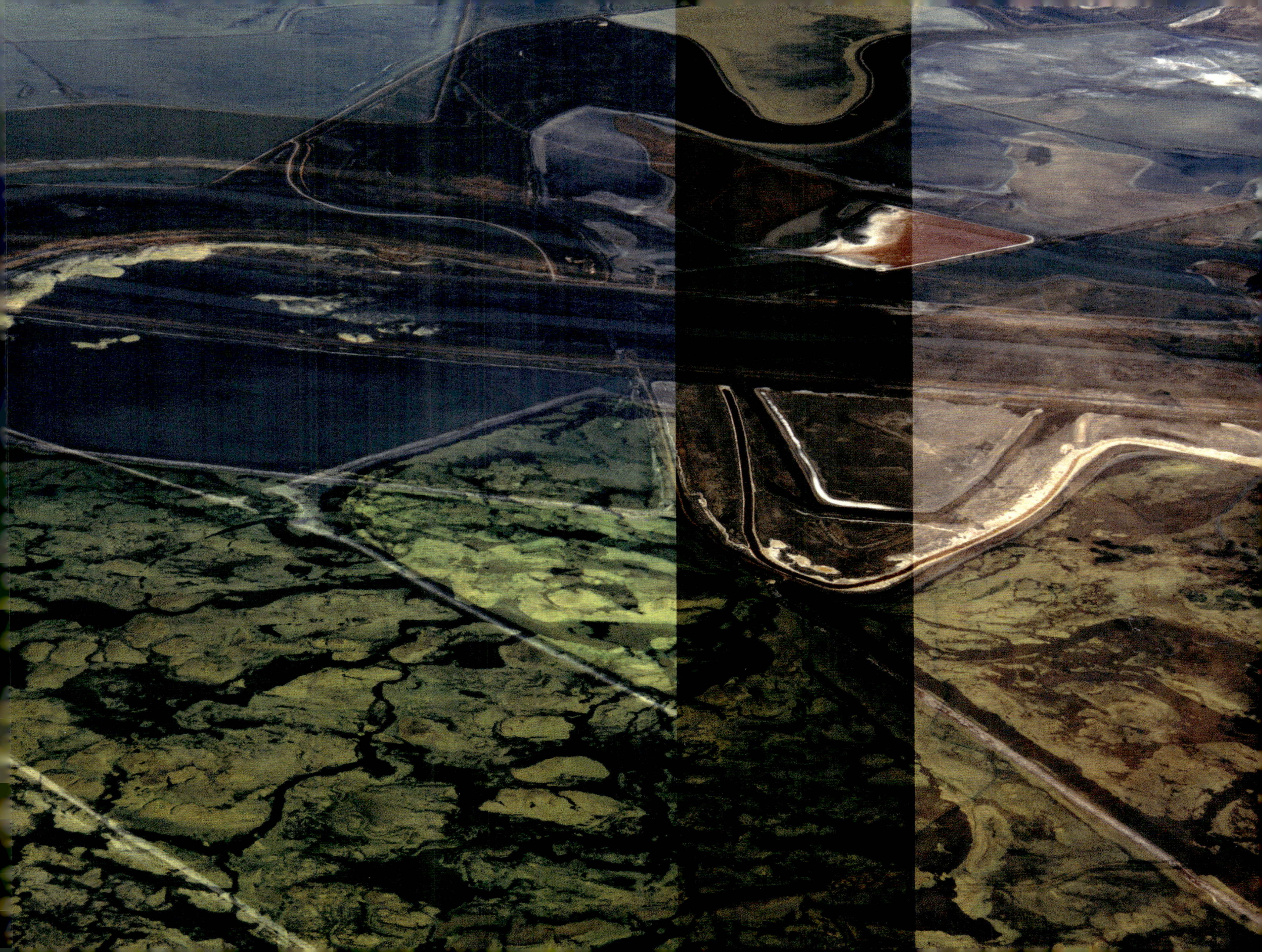

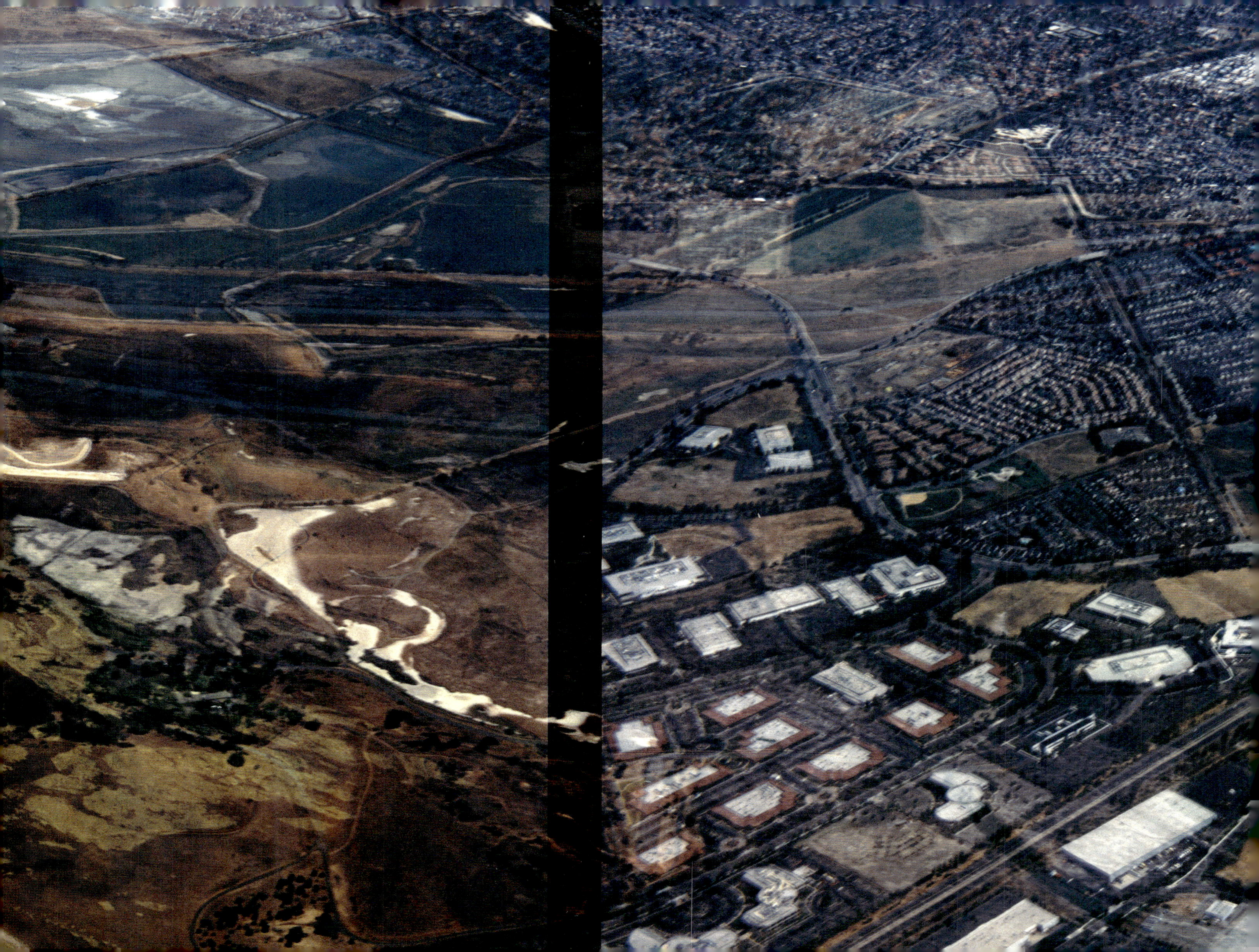

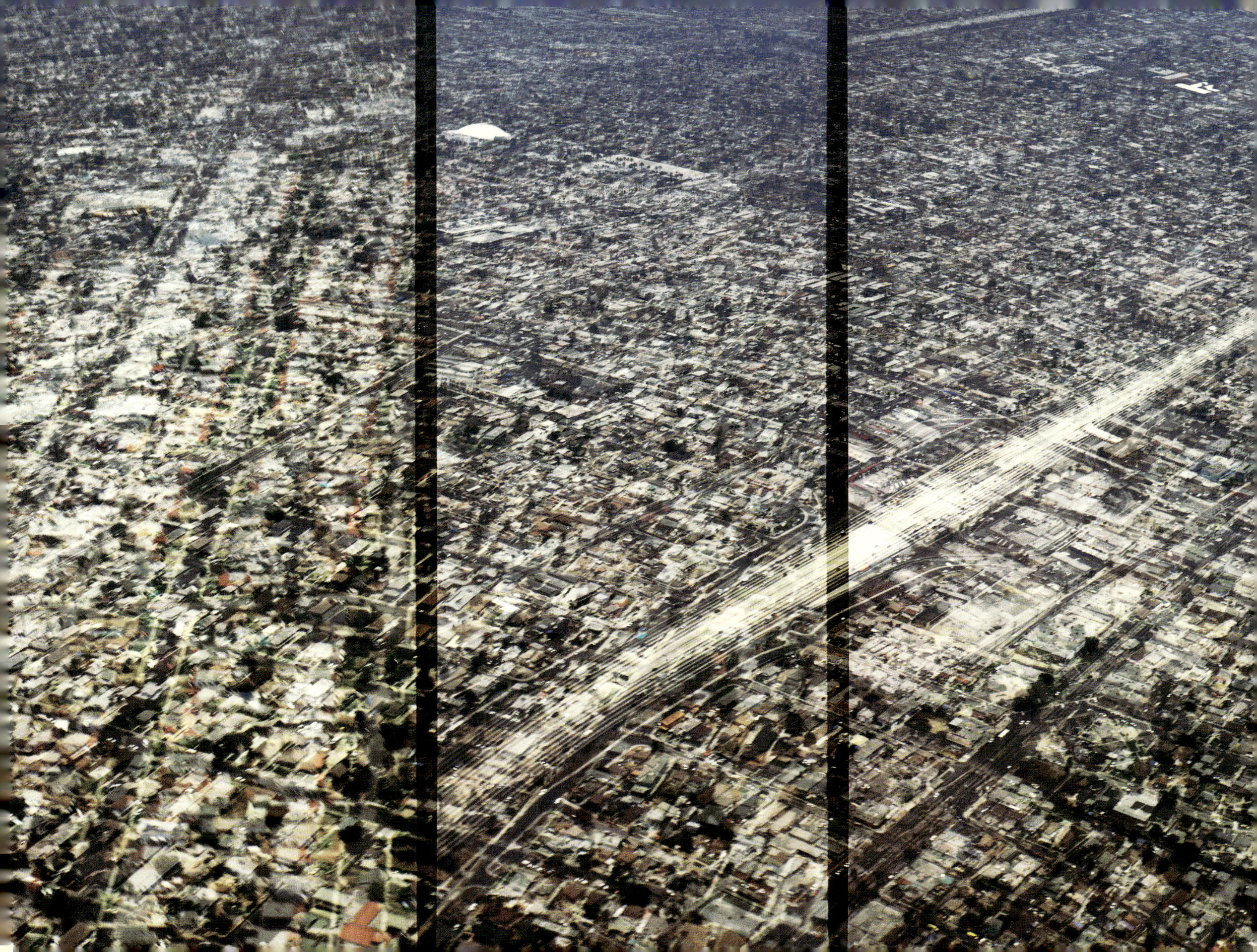

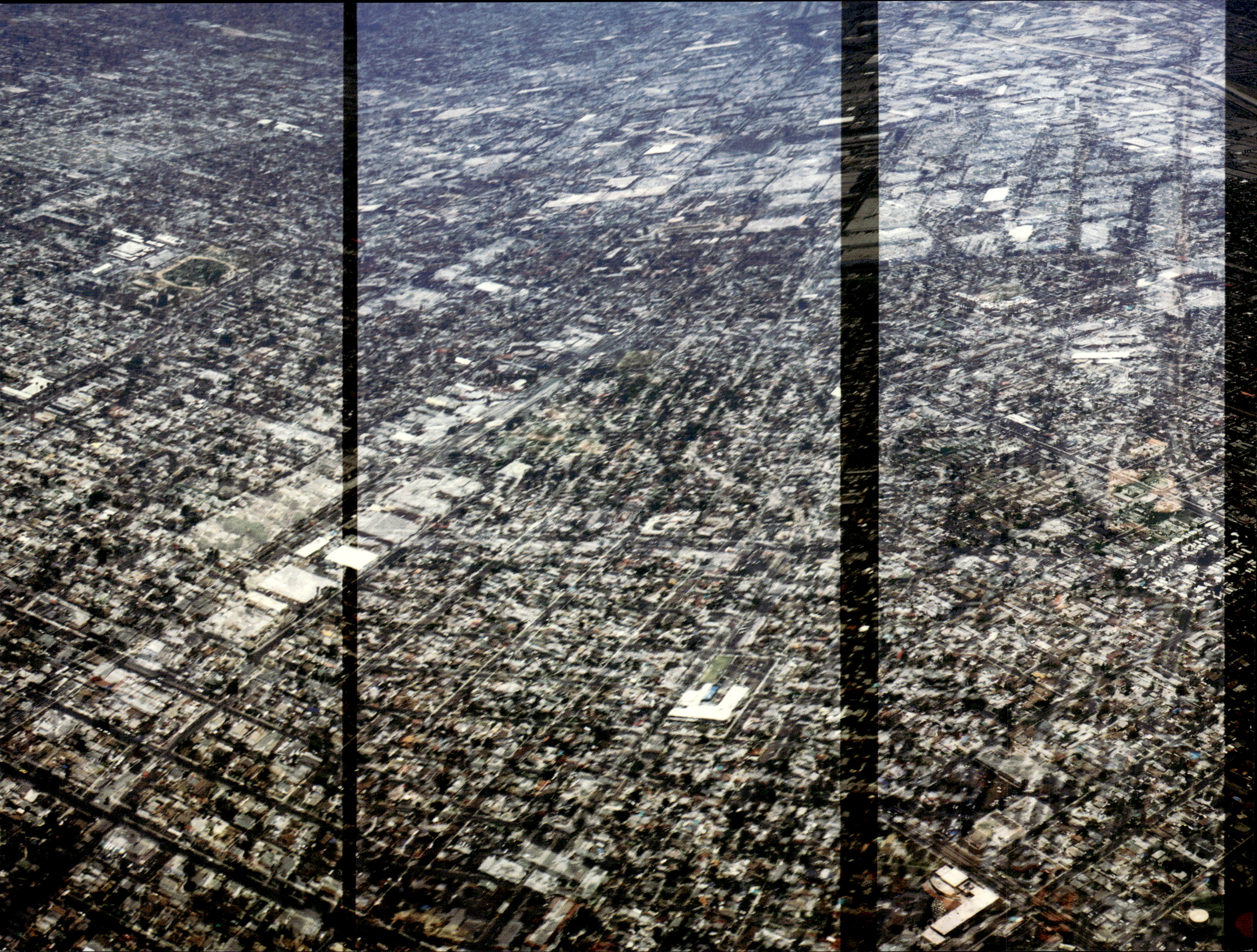

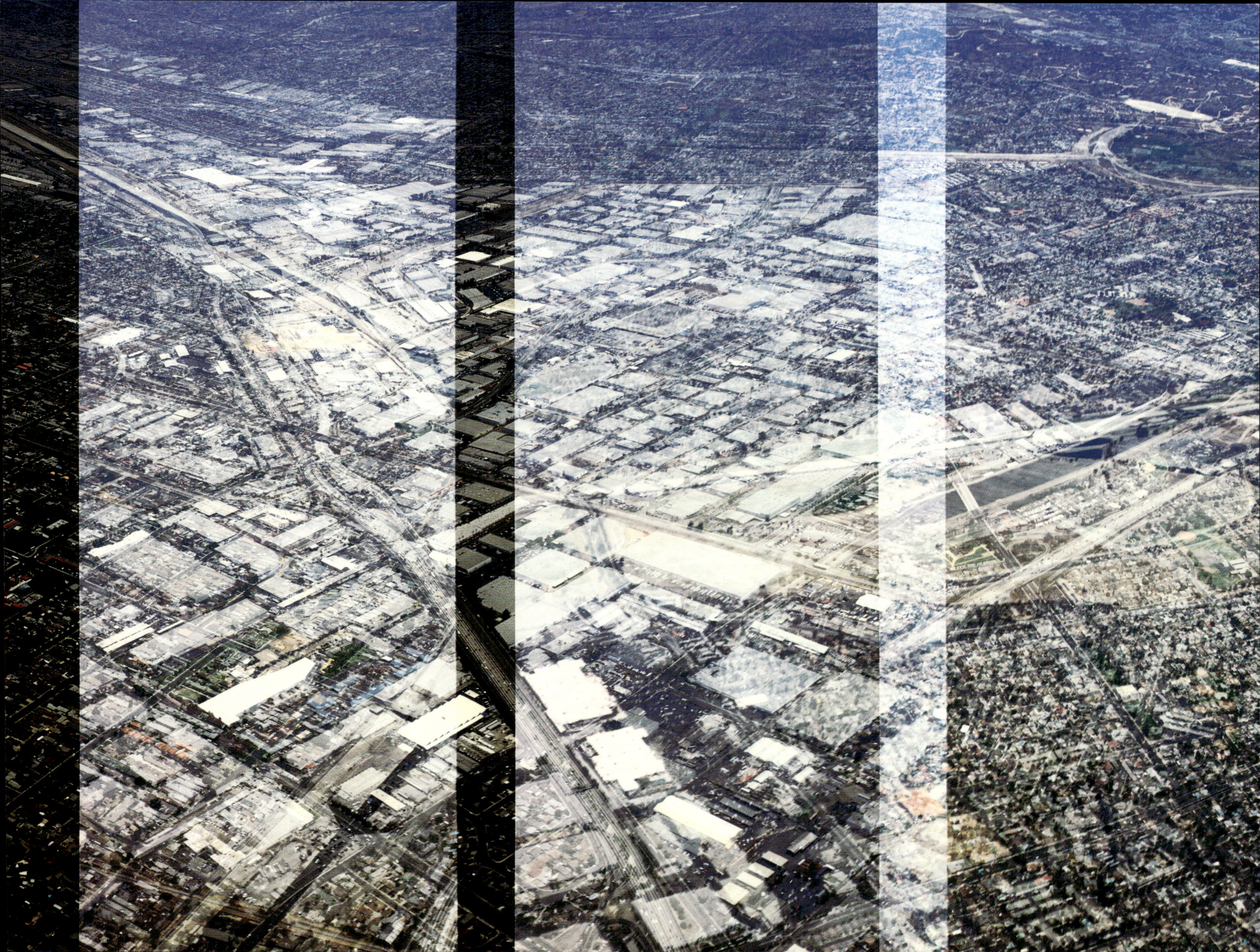

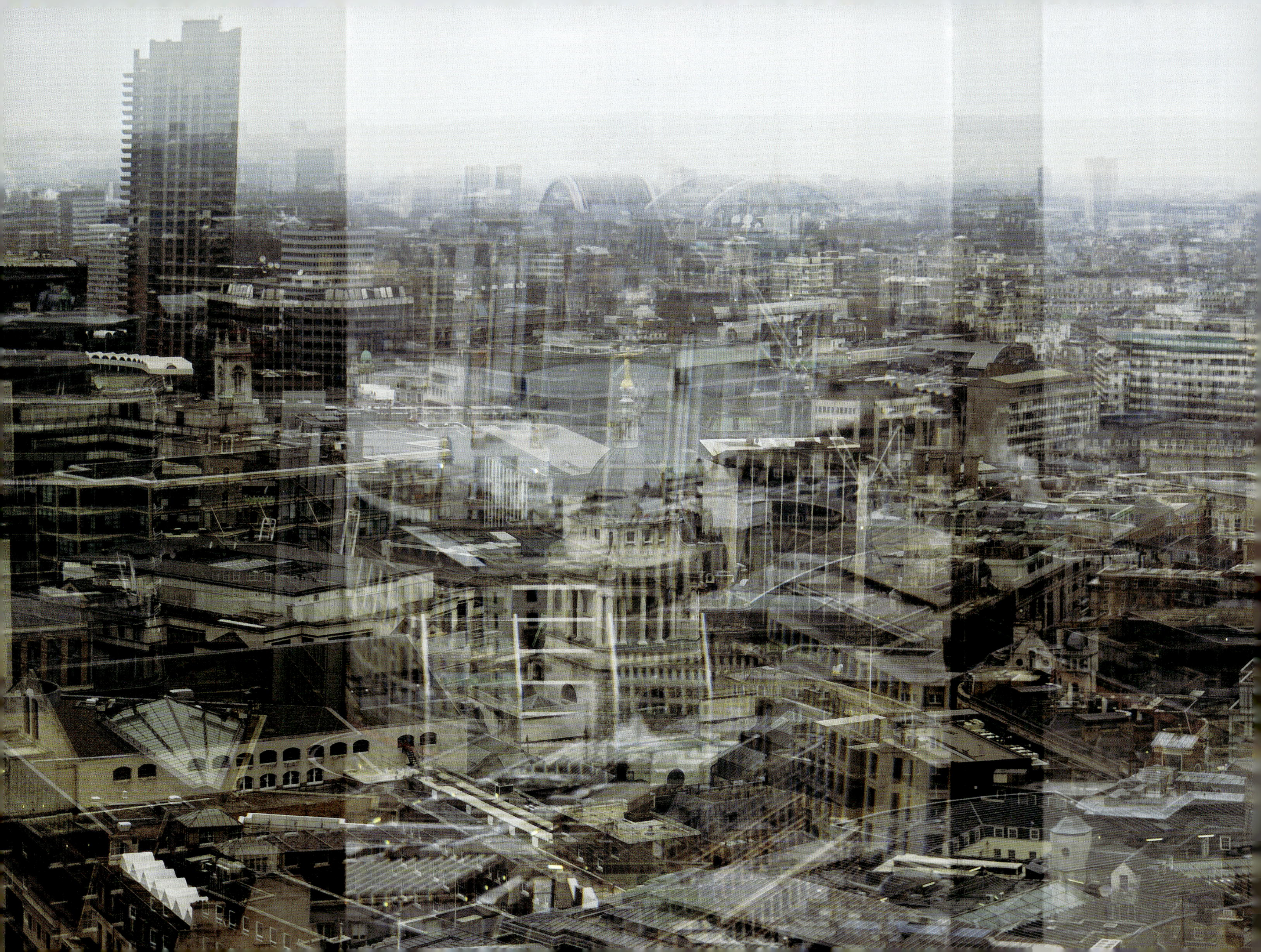

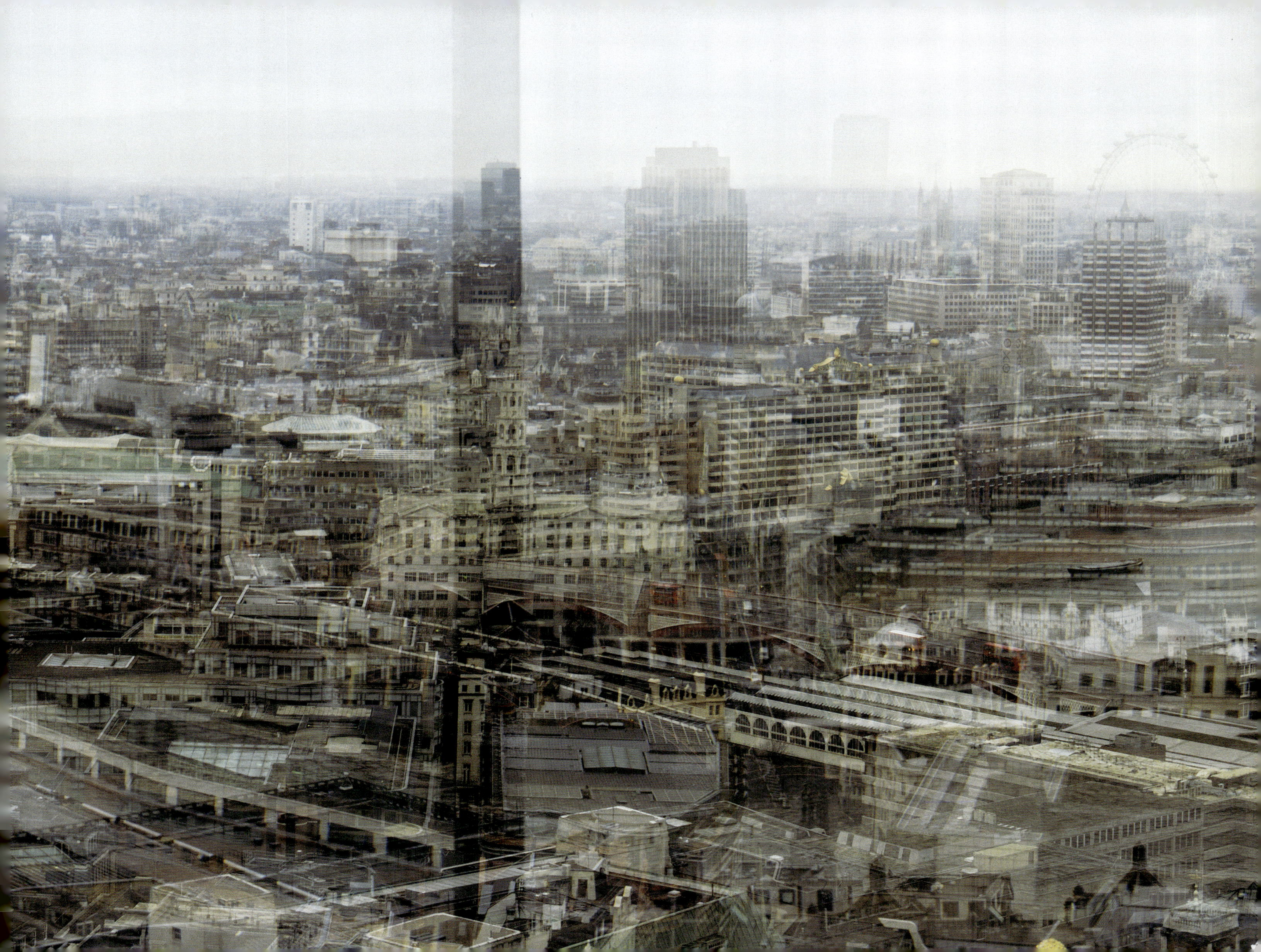

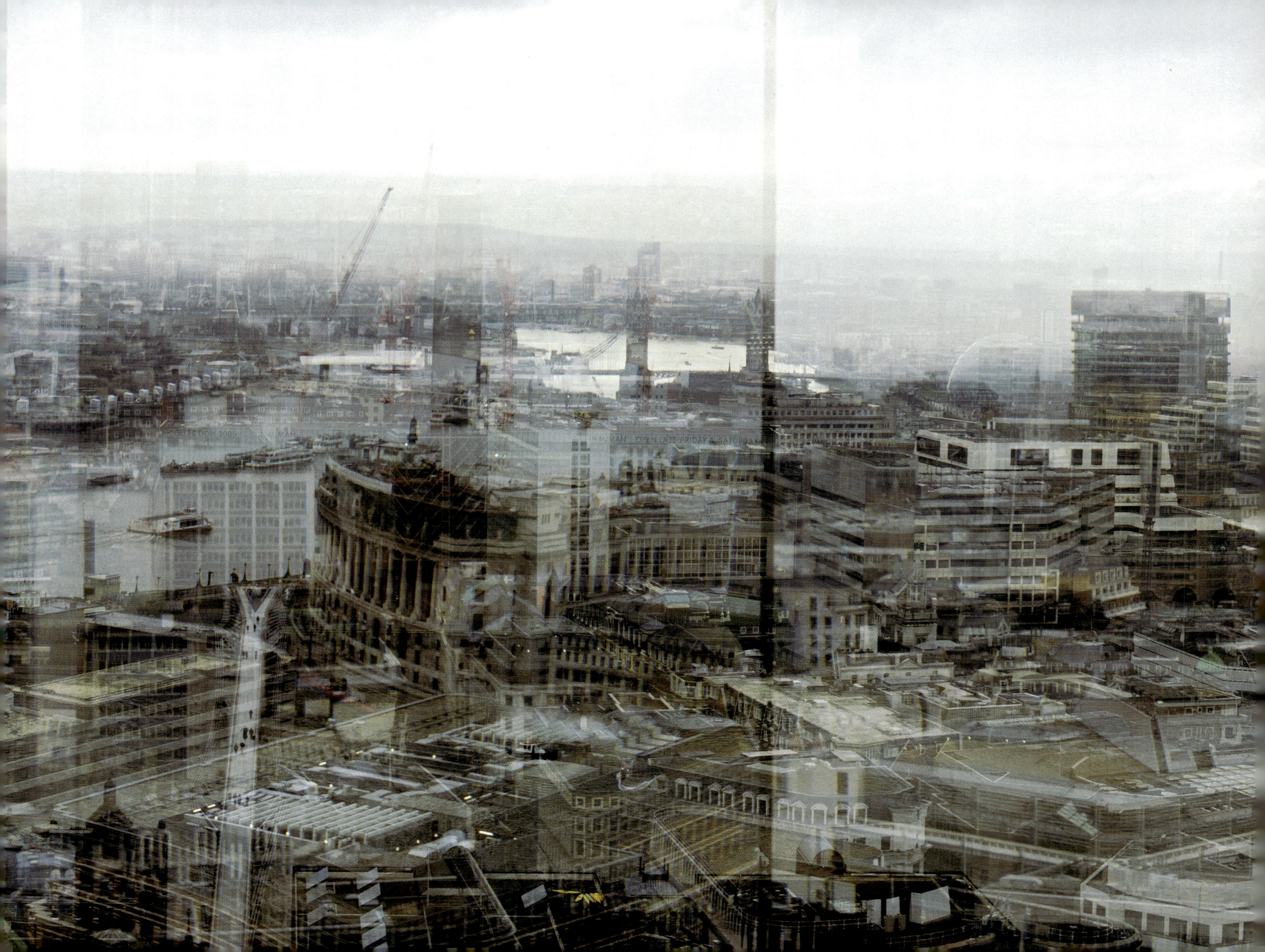

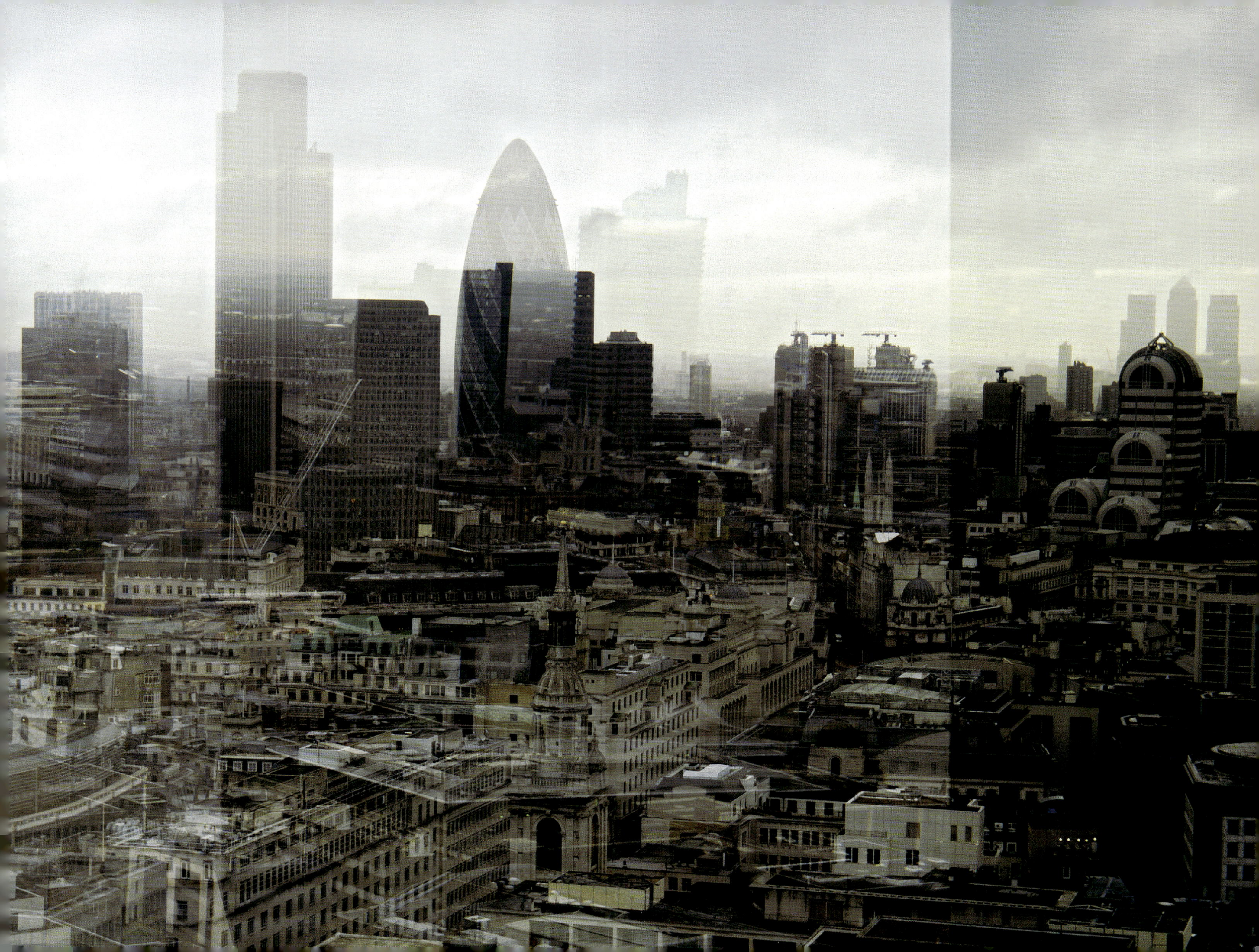

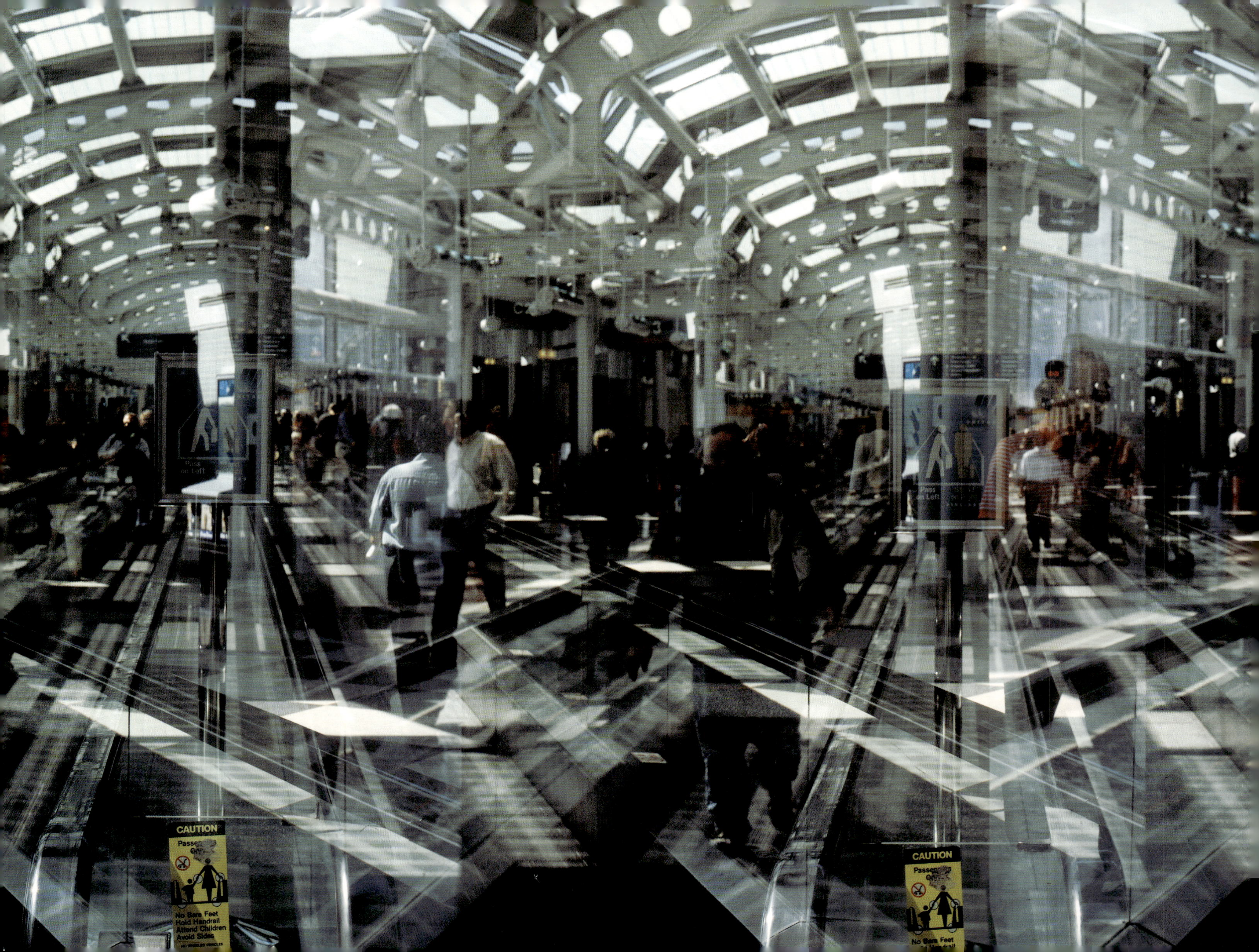
Pass on Left
CAUTION
No Bare Feet
Hold Handrail
Attend Children
Avoid Sides
CAUTION

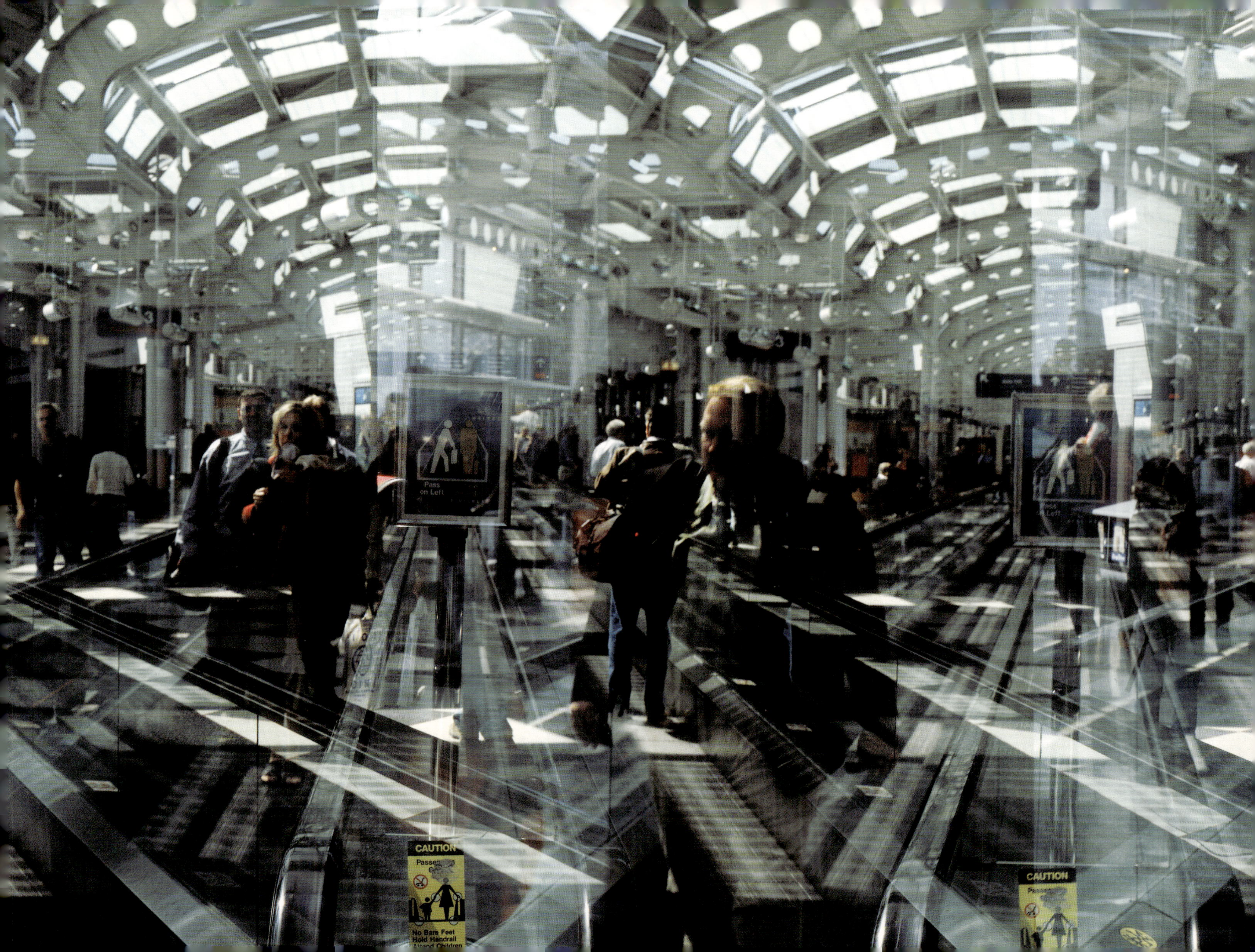
Pass
on Left
CAUTION
No Bare Feet
Hold Handrail
CAUTION

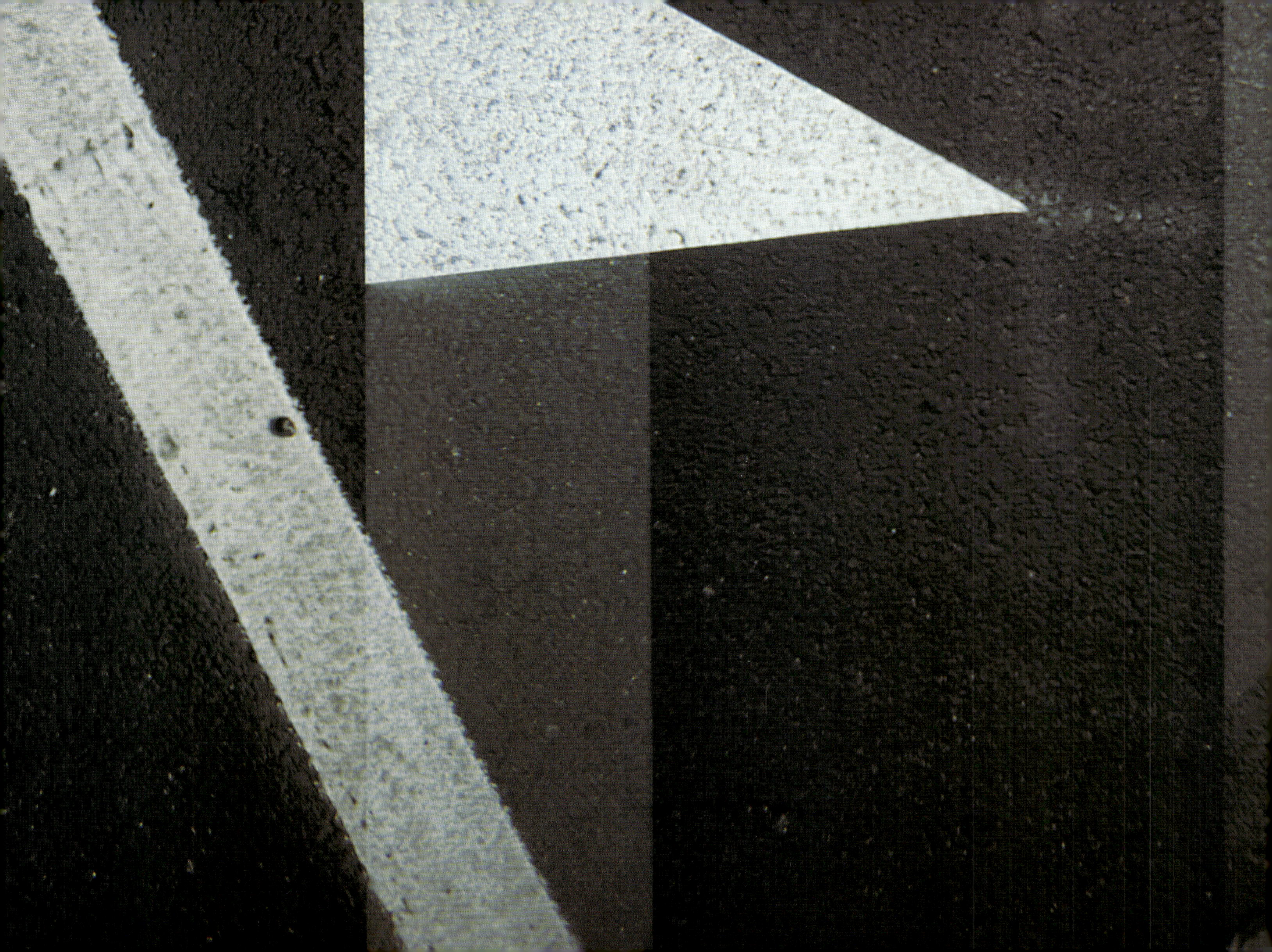

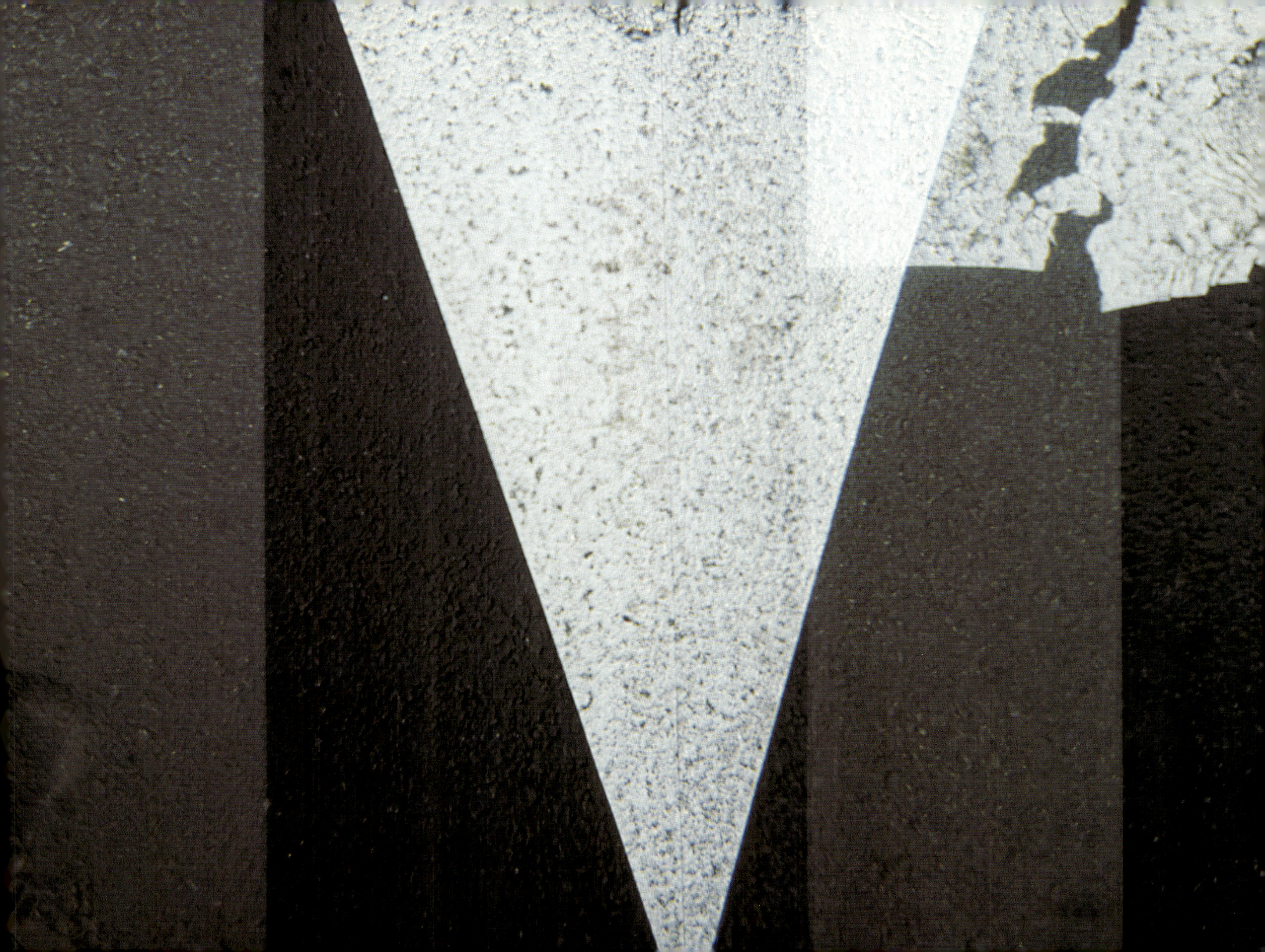

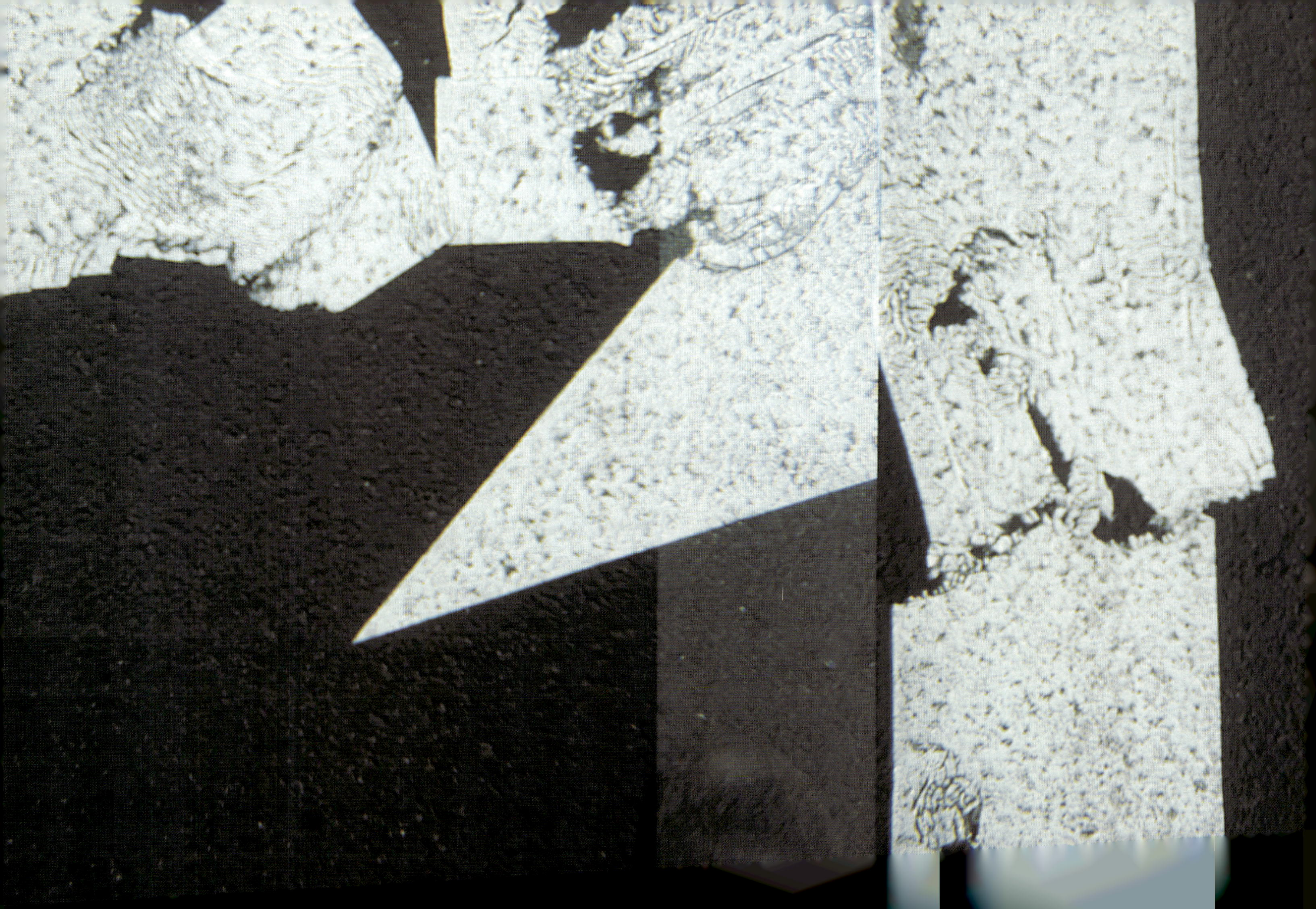

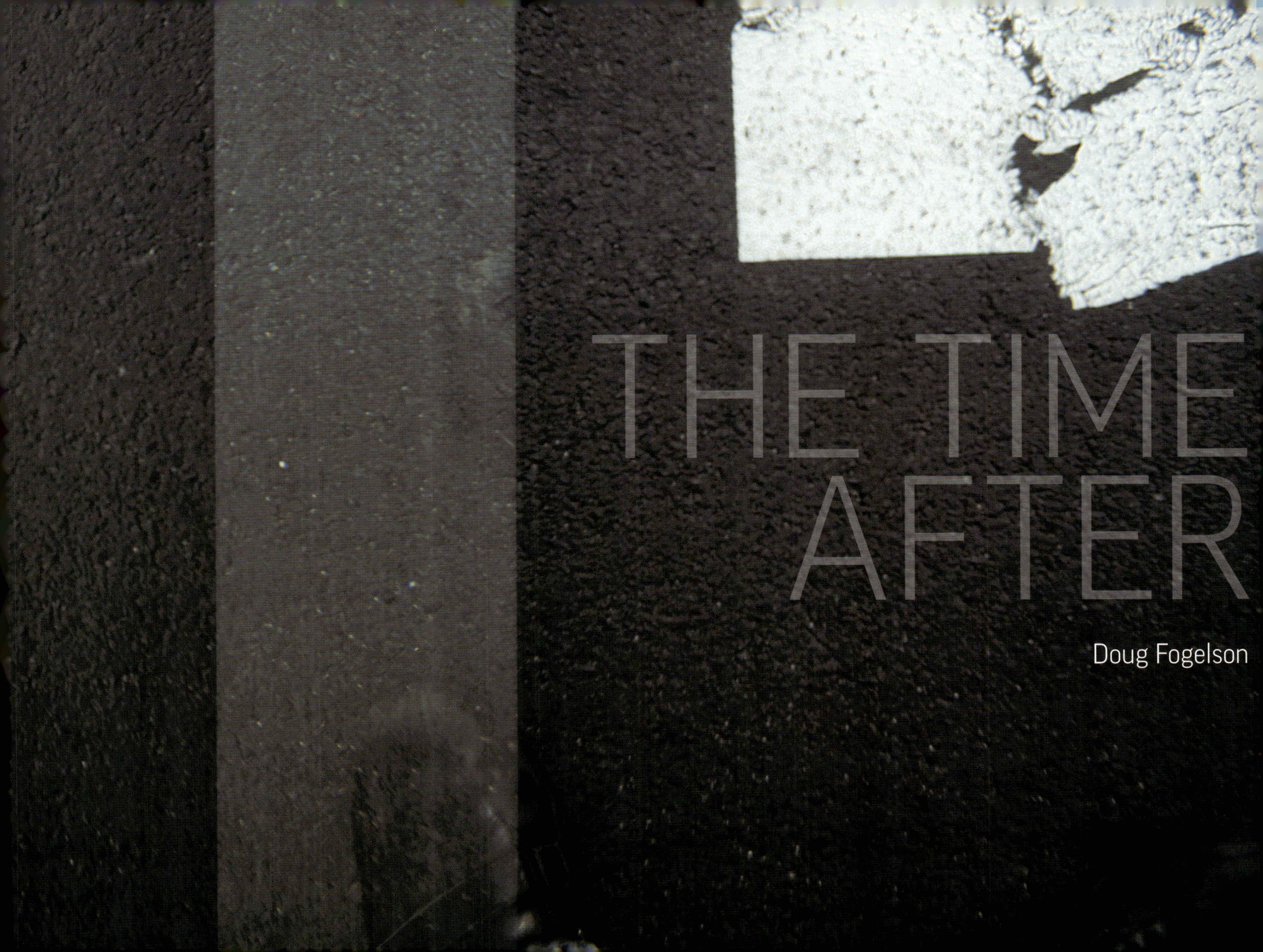

THE TIME AFTER

Doug Fogelson

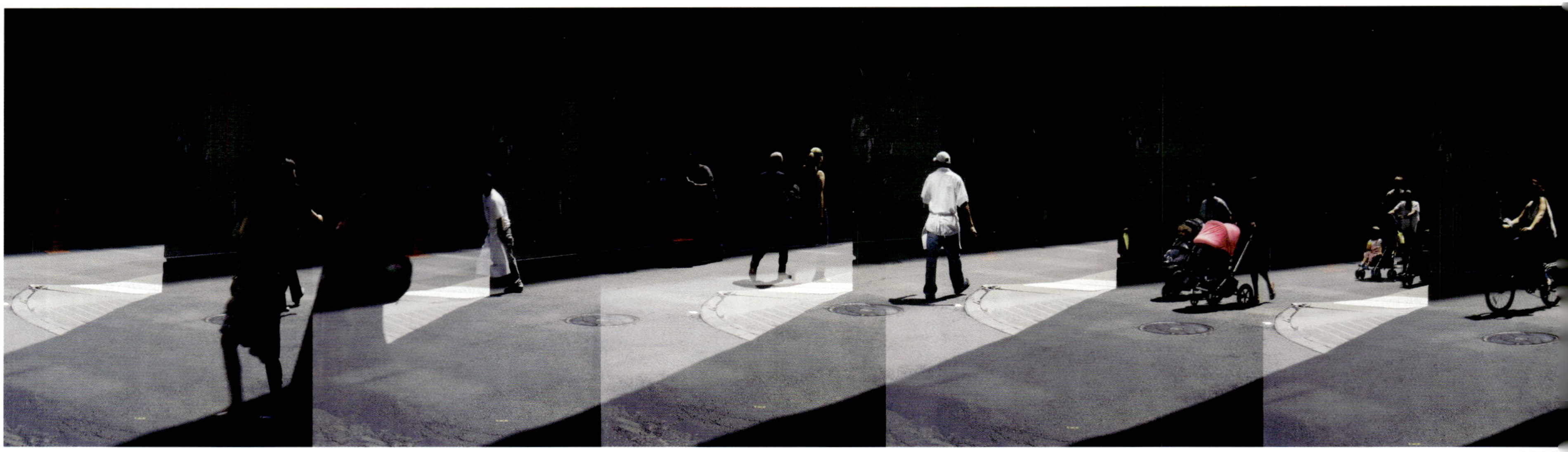

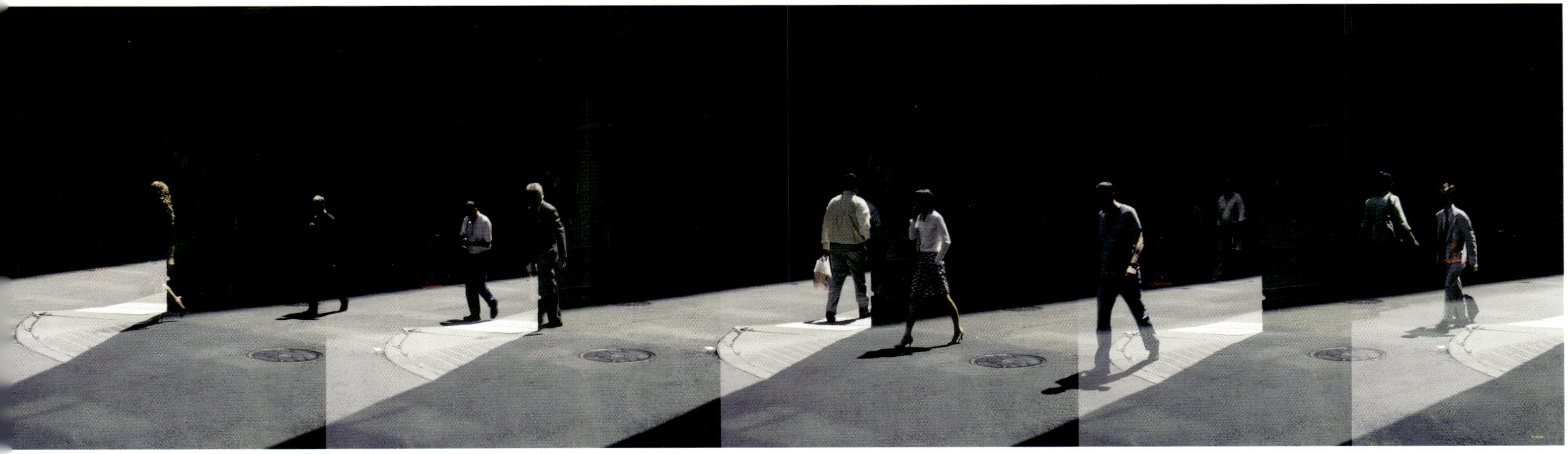

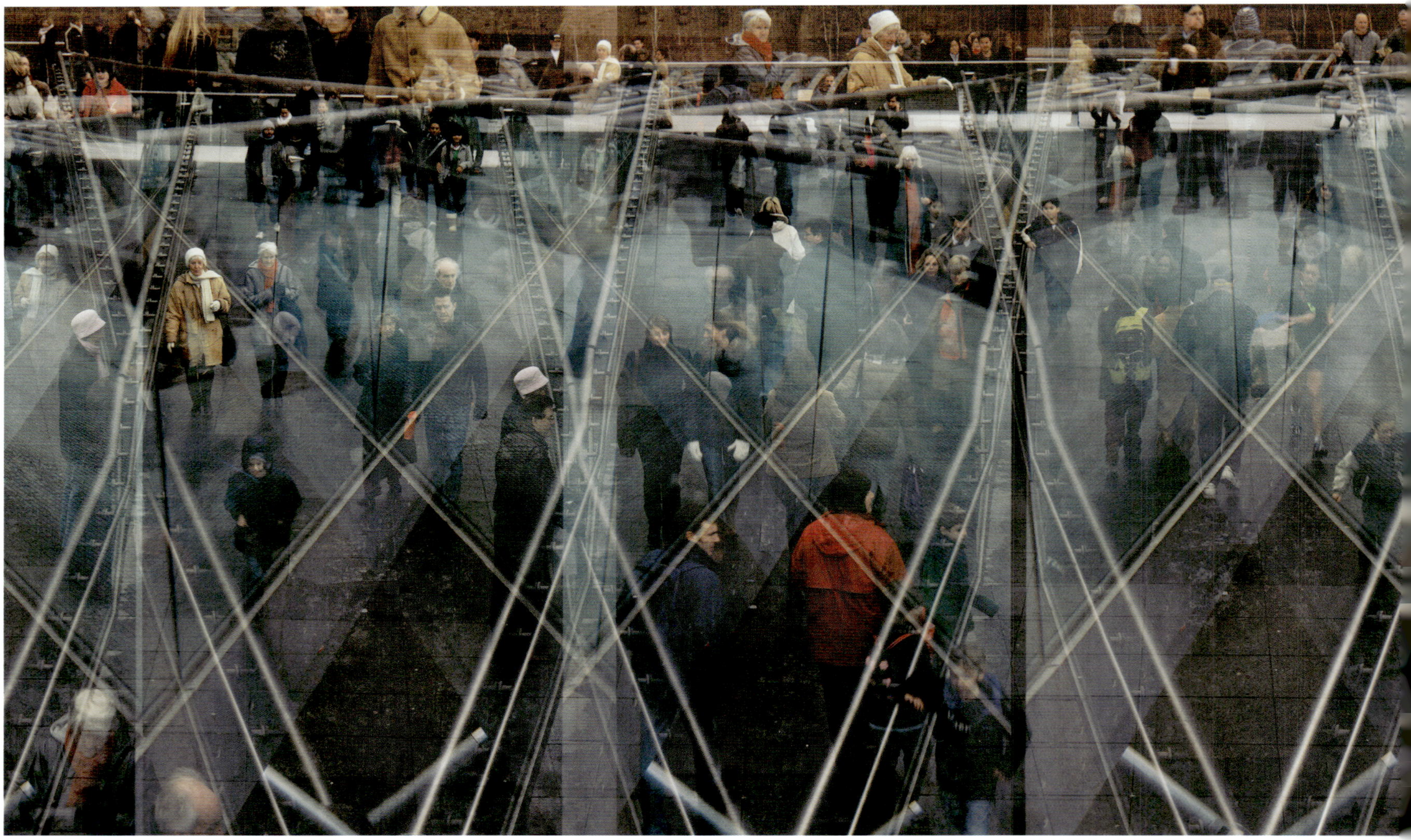

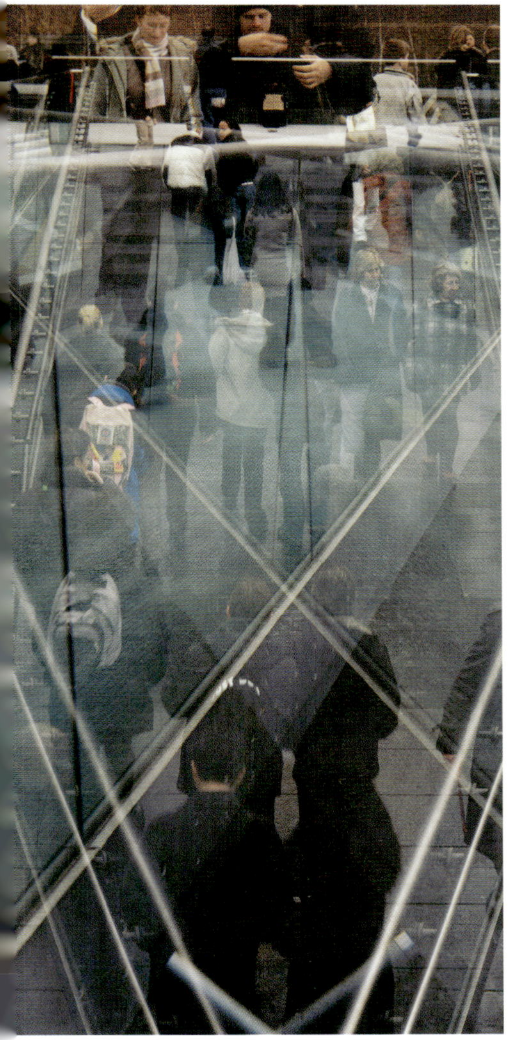

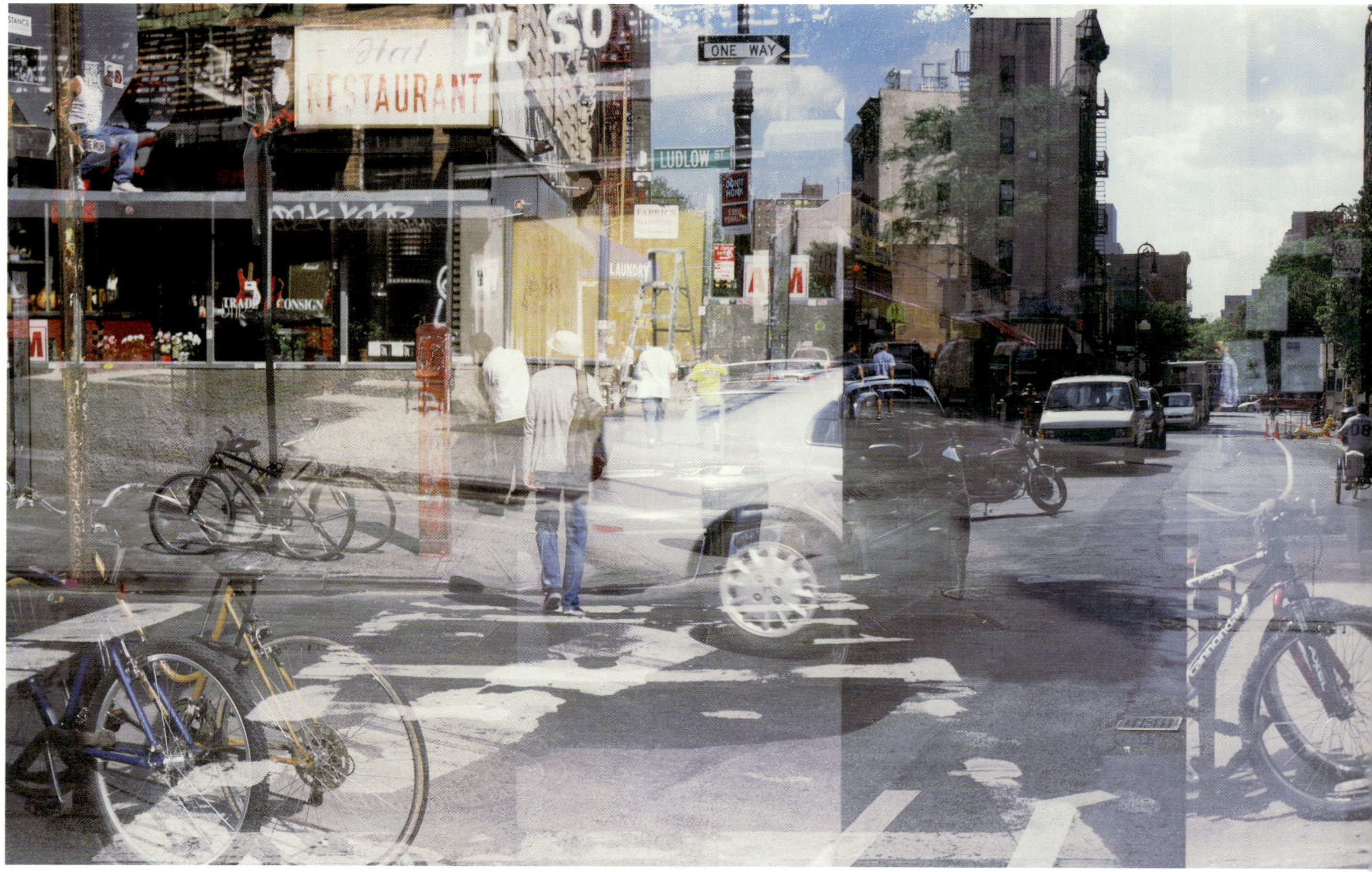
RESTAURANT
ONE WAY
LUDLOW ST
TRADE
CONSIGN
LAUNDRY

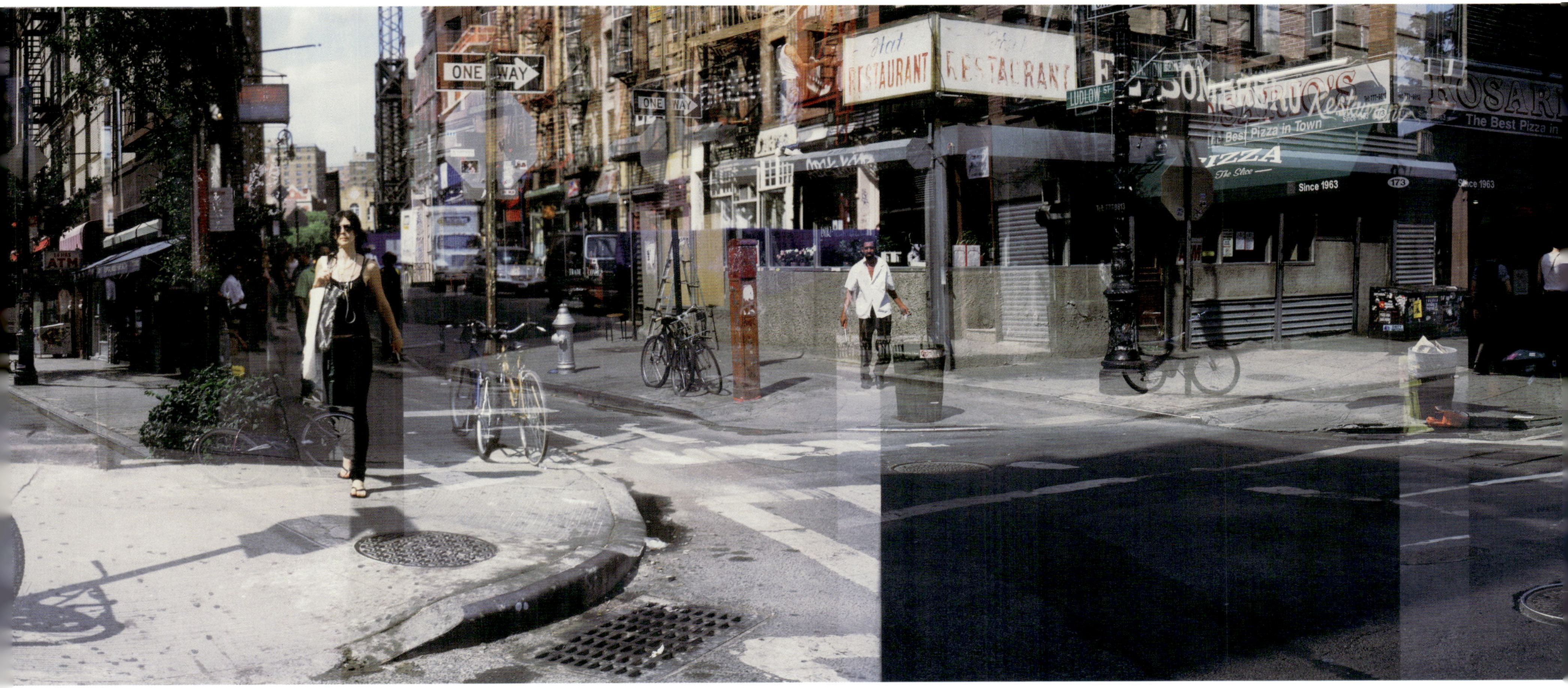
ONE WAY
RESTAURANT
LUDLOW ST
The Best Pizza in Town
Since 1963
173

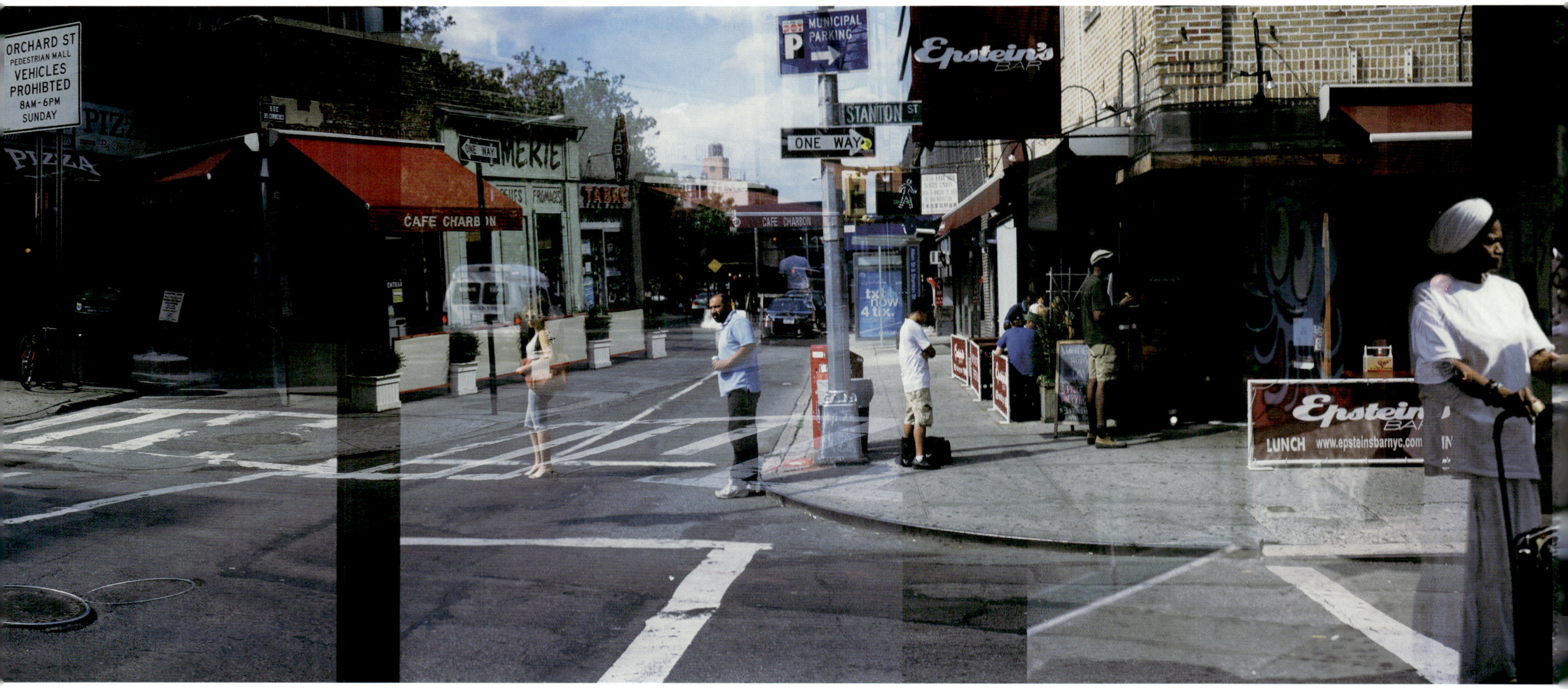
ORCHARD ST
PEDESTRIAN MALL
VEHICLES
PROHIBITED
8AM-6PM
SUNDAY
PIZZA
ONE WAY
CAFE CHARBON
FROMAGES
TABAC
MUNICIPAL
PARKING
P
STANTON ST
ONE WAY
Epstein's
BAR
CAFE CHARBON
LUNCH www.epsteinsbarnyc.com

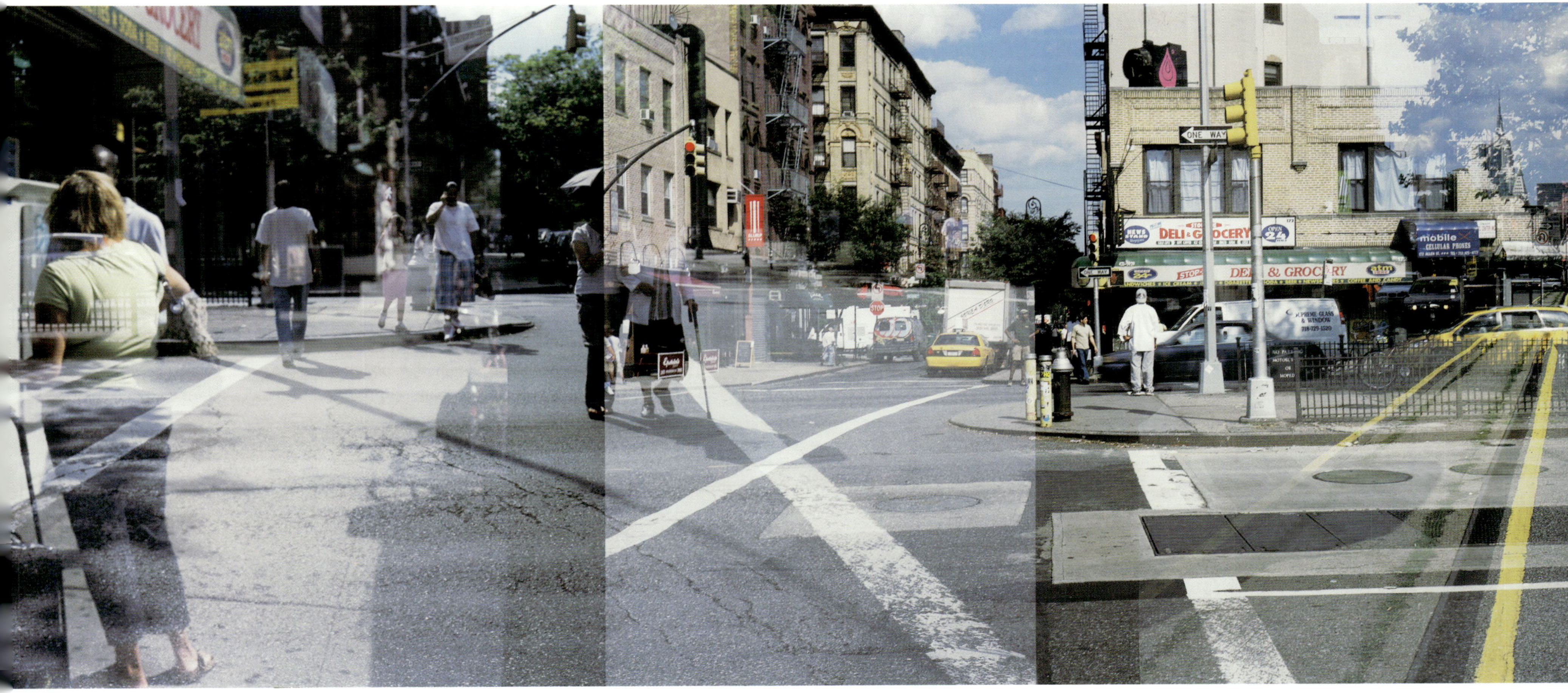
ONE WAY
DELI & GROCERY
OPEN 24
mobile
CELLULAR PHONES

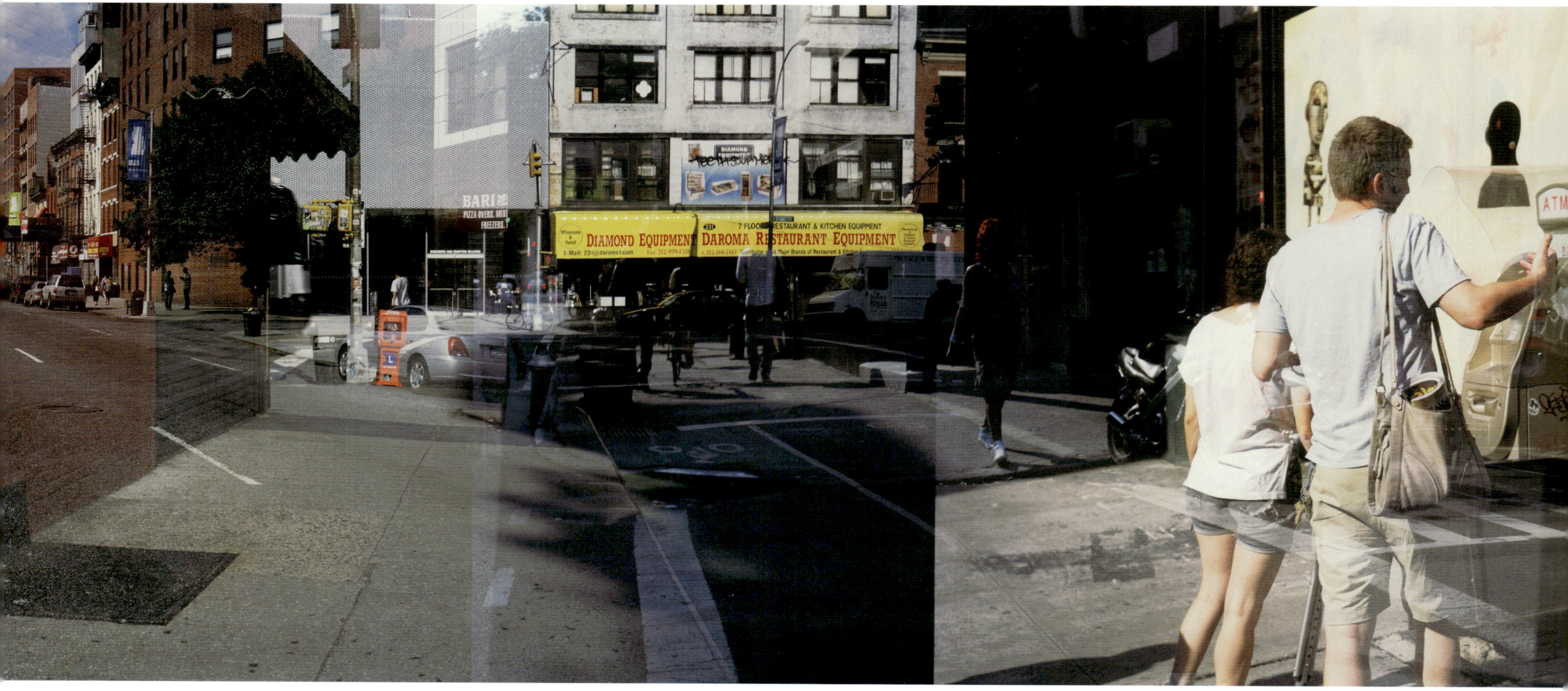
BARI
DIAMOND EQUIPMENT
DAROMA RESTAURANT EQUIPMENT
ATM

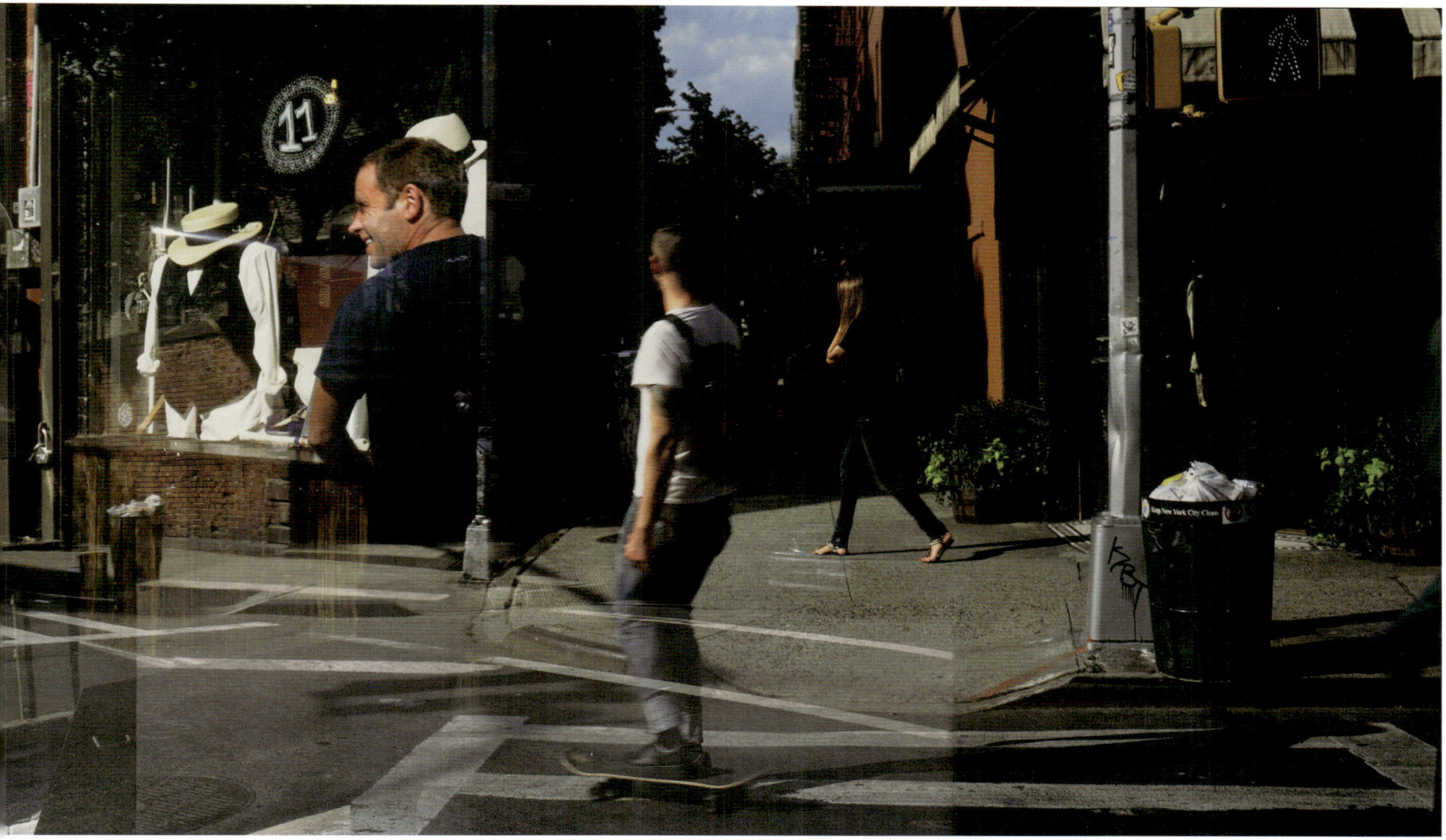
11

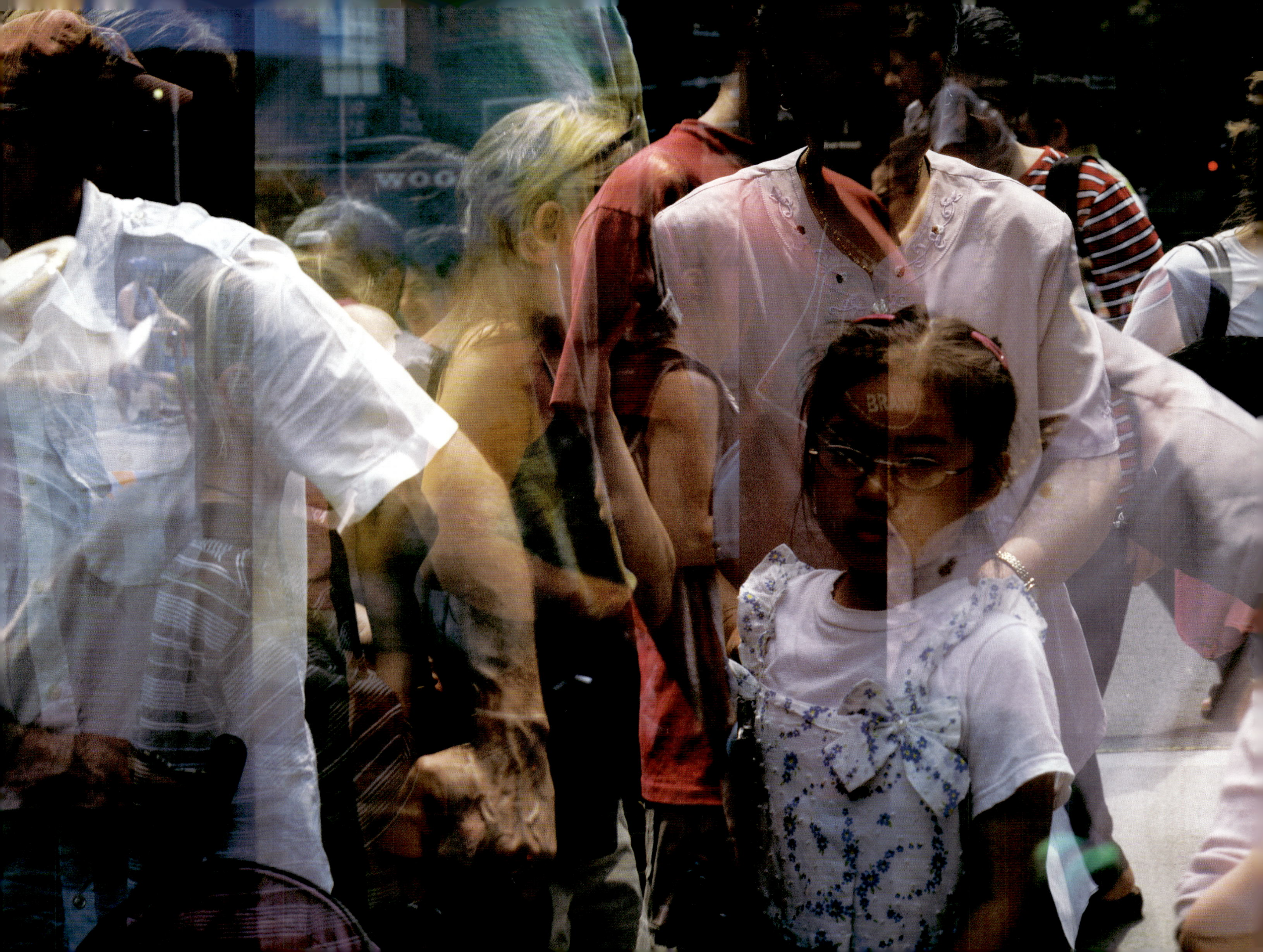

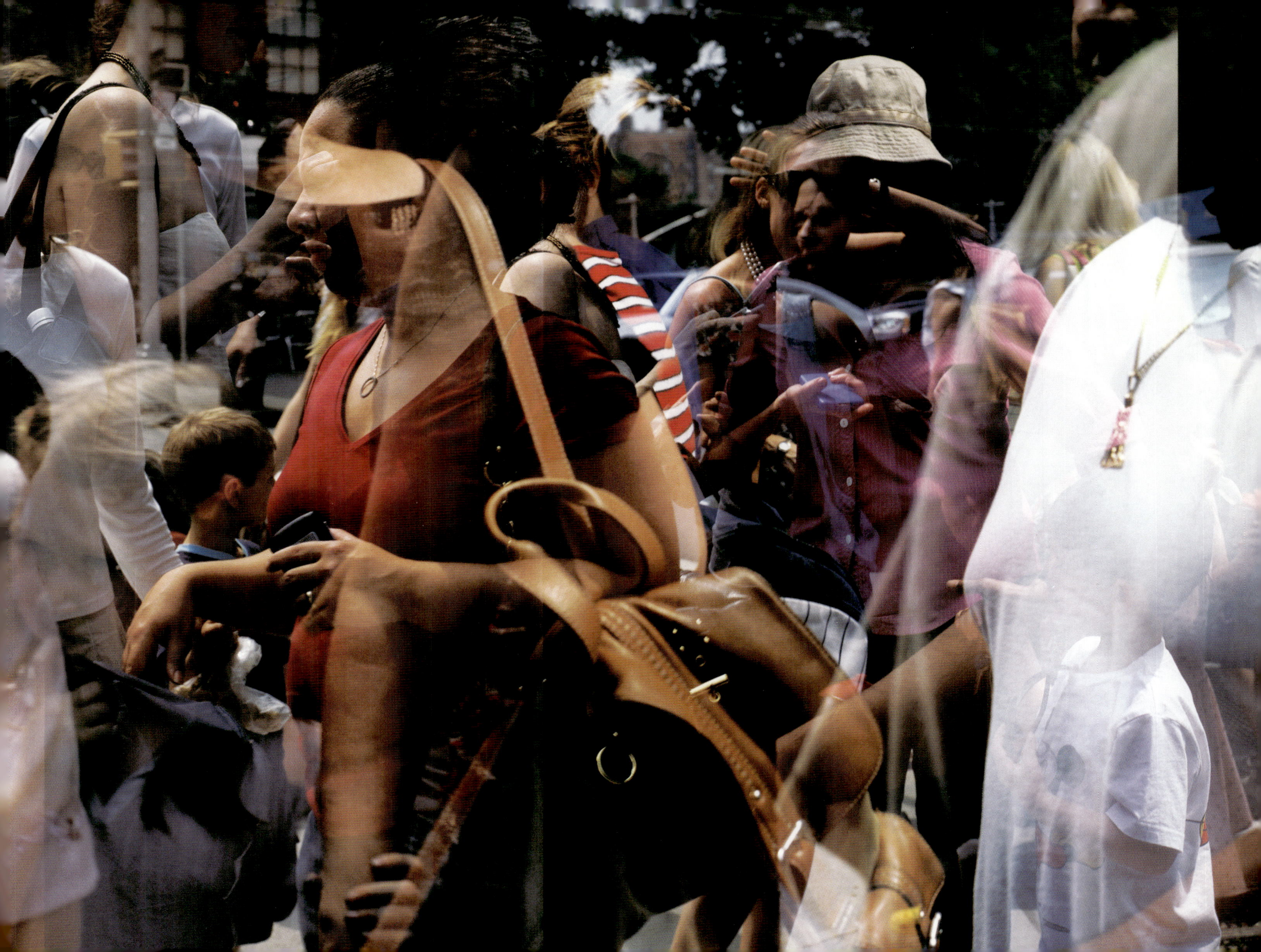

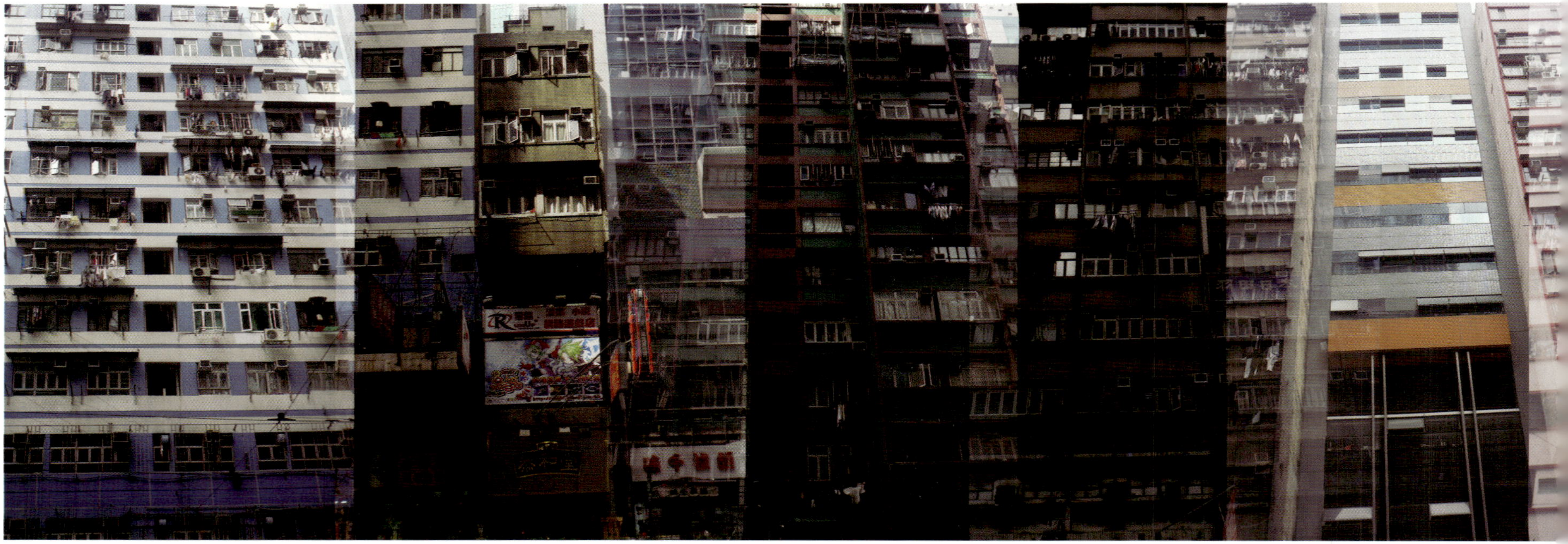

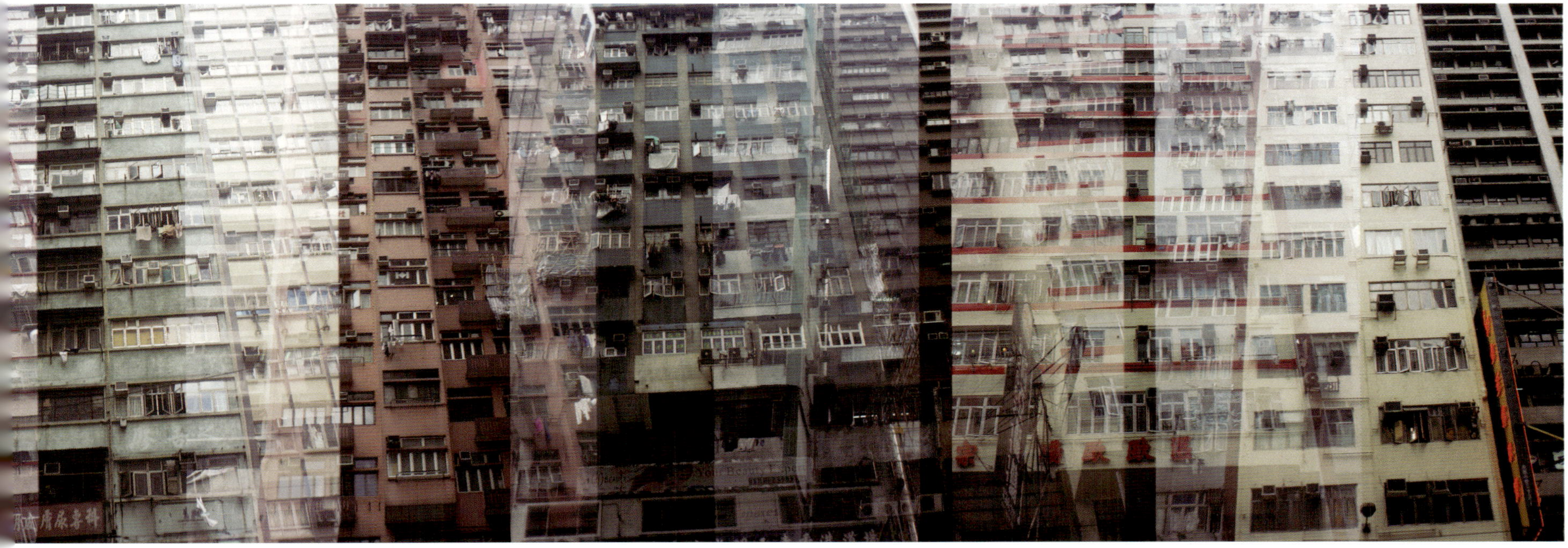

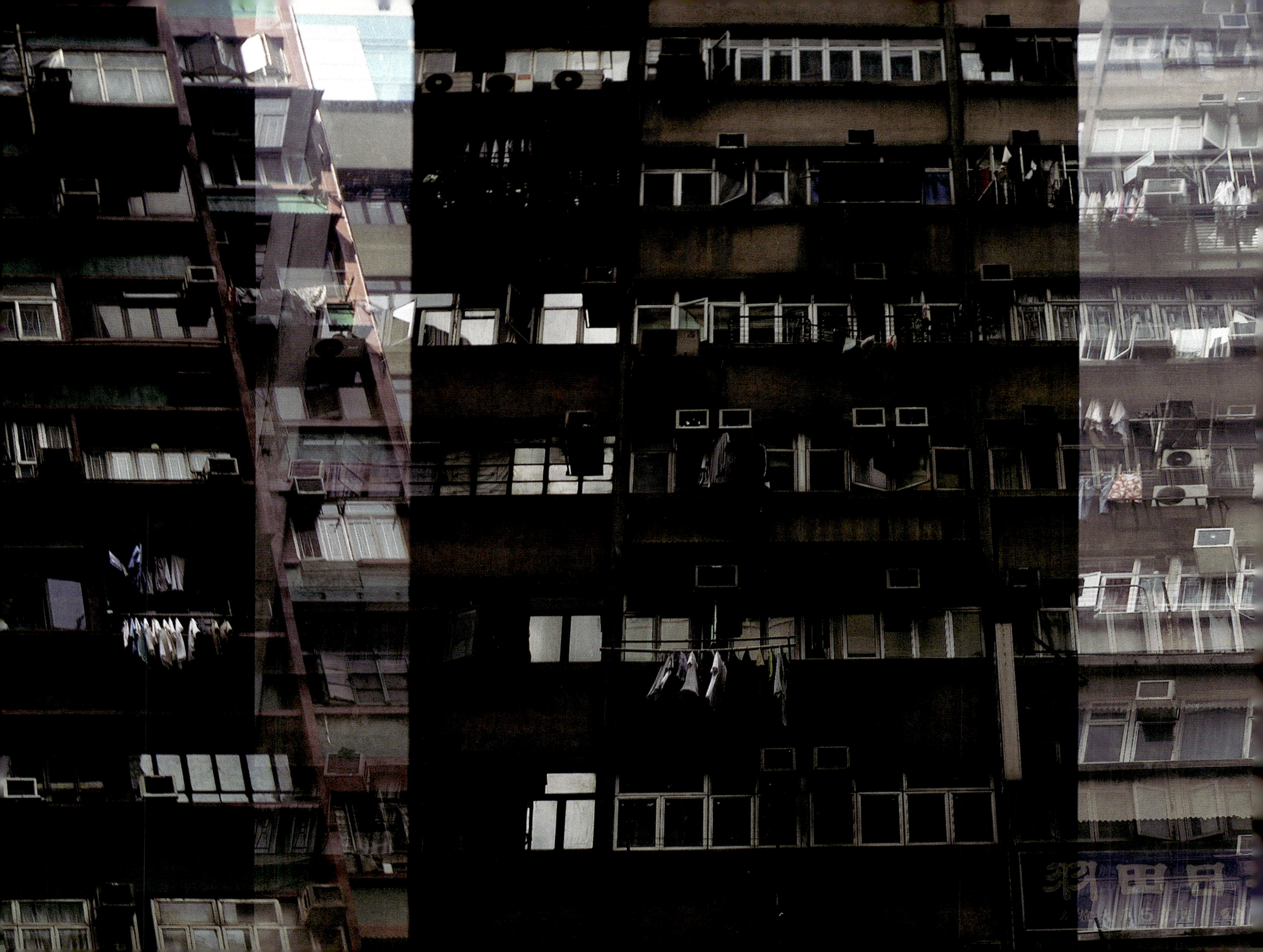

AND NOW WITNESS THE ENDING, BEAUTIFUL AND TERRIBLE.

by Eiren Caffall

PART ONE:

The Distance Between the Loved and the Lost

It has taken us so little time to warm the earth beyond repair with the breath from our factories and cars, our coal plants, and our six billion sets of lungs. It has taken us so little time to warm it to its ruin.

And even now, as you read this, the earth *is* being ruined and remade, altered so finally that soon we will not recognize it as the earth we knew. Now, even if we stop some part of this destruction, every single thing will be materially altered from this moment forward until well after human time.

You do not need me to tell you that things have already fallen apart. The mysterious and strange, the broken and unwelcome visit us every day, whether we are awake enough to notice or not. Autumn stays so late, summer gets so hot, and the birds return so soon that we've hardly missed them. That things are wrong is something we can smell like other animals can scent an earthquake.

I am no scientist. So I won't recount the facts for you. They are everywhere and you may read them in any place you choose. You may already know them better than I, in fact. But, even without knowing all the science and the math of the changes, I feel them. They rise in the back of some lizard brain of mine with a simple notion that, no, things are not right anymore.

Even knowing the little I do makes my skin ache at night with a grief that I can't cry through. If I started crying for our losses I wouldn't know when to stop. There are so many. There are so many still to come.

For all of us, the swings of recognition and regret come harder and faster these days. They come along with the hope that the changes we see will all just go away, and reset to what we recognize and long for, so we won't have to witness what comes next.

This is the time when every day I look around me and note: *today is like before—I can rest in this* or *today is very odd.* The first kind of day I can bear, the second I cannot.

I cannot bear the time I live in, because I live in the time after. Not in the time after the earth was pristine. That earth was gone long before my birth. No, I live in the time after the time when we had no idea that we'd ruined things for good.

In the time before, we thought the earth was fixed, that it was ours, and that the things we'd ruined could be repaired. And now we know that is not true. We have done things to this place we love that we can never make right. We are losing what we knew, and the end, as it arrives, is beautiful and terrible.

It is an end to what came before in its most basic forms: weather and memory, peace and seasons, migrations, habits, returns and predictability, comfort and ignorance.

And how are any of us to bear it?

Because, make no mistake, here we are speaking of death: the death of the predictable, the death of the regular, the death of the familiar and stable, the known. We are here, perched on our deathbed, viewing at last the distance between the loved and the lost.

Now is the time for a deathbed conversion. We are at the hour of our loss, looking at our god for reckoning and redemption. We would like to wrestle this moment and find grace, maybe forgiveness. But if we cannot wrest grace from the knowledge of what we will be losing, doesn't consciousness require that we face the void with open eyes?

The only thing that can help at all is seeing, seeing fully how things are now, and how they were before, and how they may become. Truly seeing in this time after may give you strength to act. If it gives you no comfort, at least it may give you the hope to change, the power to fight, the peace to endure, the patience to teach.

Now is the time to see what you have loved and what you have to lose.

PART TWO:

The Time Before Is Dead, Long Live the Time Before

So, first imagine you have died. Trust me, it will help.

It can't be the first time you have imagined this. We all have a language for it, whether we speak it often or not. Romantic, nostalgic, or heartbroken, these are places we have walked before.

Cast yourself into that place now, and think of how you would mourn what had passed after your own life ended. Think of that proverbial view back to the planet from the place beyond. What do you see? The earth and all the people that you loved go on without you, they live and thrive and mourn you.

And what is the loss you feel as you look back? Do you mourn the specific places you can't walk anymore, the things you can't touch? Or do you instead mourn the things you can no longer affect, the conversations you can no longer have, the advice you can no longer give, the role of caregiver to your loved ones that you can no longer inhabit?

Because now you can't have any role as commentator on the lives you left behind. You cannot add to the conversation about the world as it is, your family as it proceeds, your work as it goes on or falls apart. You cannot leave a legacy or change a mind or alter a single thing. It is your impotence and your distance that pains the most.

Now, as you are hovering in that place, picturing the world without you, shift your focus ever so slightly and imagine instead that you are looking back at the earth of the time after.

Imagine more than that, imagine that it really is too late. The eleventh hour has come and gone, and we have missed it. The wreck we've made of this culture of ours is not only permanent, it is catastrophic. Imagine it is all gone, everything we've been, all we've built and protected and fought over, along with the hope of ever making it right.

The places that you've loved are gone. That stretch of beach in Maine is under water. That park on Lake Michigan now extends across acres of new land to an anemic little patch of acrid water, choked with weeds. The mountain where you once went skiing is warm and wet in winter, free of snow. That city where you grew up has washed away, empty buildings echo and rot, covered in mold.

Imagine that it is worse than that. Touch your worst fears, perhaps only with the edges of your mind. The waters have risen, the storms are ferocious, growing patterns and climate and access to food and resources have been devastated. A few thousand of us are left at the poles, living on the only arable land, far away from a culture now swamped by risen seas and global war. Or maybe we are all gone. Push the borders of your imagination out as far as you can stand.

Now, here you are, in that place beyond, looking back in longing at the world you can no longer touch. What do you miss the most of what was lost? The forests and their quiet light? The drip of water off the palm trees during monsoon? The way an olive tree lent shade to the rocky ground? The color of sunlight over the Turkish river valley as spring swelled buds on the plane trees? The hat of snow topping a purple crocus? Rows of corn bristling away to the horizon while the sounds of cicadas fill your ears?

Or do you miss instead the work you did to keep all those things remembered, keep them studied and safe and recorded. Do you miss the power to affect them, protect them, or revere them? Do you miss your own imprint on the earth?

If the worst happens, what will go is our control, or our sense of it. The earth will function without us, of course. Most likely there will still be forests and streams and flowers.

The loss will be that we will no longer be here to witness and record them, to love them. For in the end, isn't that what we are here for most, more than survival, more than any thought or art that we can add?

We're here to see it, to know and feel the earth, and be the witnesses to all it does. So, to be ready for survival we must first witness the worst thing that may come to pass, if only in our imaginations. We must witness the loss, the ending, in all its glory. Because without losing everything, even in your mind, you cannot hope to win it back again.

PART THREE:

The Place You Used to Love Goes On

So, how do you win it back again? You win it back by remembering that you can love it no matter what. You can love it even if you cannot be with it. You can love it even if it is not the same. Like life itself, you can love it even if you don't have it anymore.

Because let me tell you, what is coming may be sad for us, but for the earth itself it is no tragedy. It is nothing that she cannot recover from in time. Nothing we've done compares to years of tectonic struggle, asteroids, volcanoes. We have done something unnatural, but it is all just a little thing to her, and she can go on without us. That is her beauty.

If you could perch, benevolent and uninvolved, over the earth, imagine what you'd feel if you could simply watch the changes unfold.

Start with the sky. Picture the clouds, like curtains over the face of your beloved. Between you and the planet, they hover, rich and heady, thick with stacked-up vapor.

Everything we've done is in those clouds, every breath on every breath. Every sigh and sip of air, ragged sobs, and little gasps. Once it was a blanket, holding out the cold of space, and it was heaven. In a way it is a blanket still, but now, like a blanket in a room full of overactive radiators, it is just too much.

You could look on the sky with fear, as it has become an accomplice of all that heat. But, you are gone, and so you may be able to admire the clouds from afar again. After all, the heat may dissipate in geologic time. Admire the shapes and the shelter of the clouds again, the way that substantiality and permanence are implied and then rejected by their shifting forms.

Admire the sky as it will be when we are gone. The storms are stronger, they rake the earth and sea, they swirl with huge eyes. The ragged, devastating power of their breath rushes over the planet, like so many operatic moods.

But when a mood has passed, the clouds part, and the sky shows blue as if it has just had a good long cry. Still moist around the eyes, it looks out, over everything that got it weepy in the first place, and sends calm breath over it all.

Most likely it looks down on the sea. Most of what it sees is water after all. Look down there yourself; the sea is full of the stuff of tears and sweat. You remember her like you remember the swish of your own blood.

Did you hate her power or did you marvel at it? Rise or fall, warm or freeze, she was the thing we danced to. How much land did we have to use, how much rain, how much warmth? Lush and sweet, how much green was there for us, how much sand, how much weather and sky and storm? As we added layers to the sky with carbon that by rights should have been under the surface of the fierce ground, we added more and more to her empire. How is she in the time after?

She rises and warms. She slides a hand along the shoreline and finds erosive means to take the land back to her. She warms again and moves a wave along the length of ice that made an ice shelf. *Come back,* she says to it, *you who were once made of water. Come back from where you've gone, so cold and distant, pretending you're the land.* And finally, the ice shelf relents.

Under and all around her is the ground itself. Here was the home of all that green, here was the home of all that carbon. Here is the place we walked and moved. It is harder to fear, easier to miss, since the specificity of place was more familiar to us all.

Maybe you cannot imagine the land itself in the time after we go. Maybe you wonder how the forests fare, how large the deserts are, how much ice there is, how the grasses and the fields, flowers and the gardens go on or fade away in the absence of us.

Well, of course, some places grow that never grew before. Some are robbed of water that they needed to survive. Some are under water; some erode at a pace that makes them gasp. Crazy flowers bloom in the Arctic. Odd and lovely jungles sprout. Trees come back, those holy and necessary giants, and claim the space we kept them from.

At the edges of the land, and perched atop it, are the cities that we made, the places that we changed. Are there any of us to remember what they were for? Are the places of commerce and culture of use to anyone or anything, besides the feral cats that walk the empty buildings? Are there ghosts of us, hovering over the places where we once made ourselves busy and useful, prescient and vital?

The planet must echo with our ghosts, and with all the things we did and wanted still to do. Most of all, it echoes in our absence with what we couldn't do to save our role here. It echoes with our failure to save ourselves, to make ourselves important to the future of the place.

I know I fear all this, as you must, too. And yet, if you can remove the terror, the grief and the guilt, the uncertainty and the self-delusion, maybe beneath it all is the simple honest love we've always felt for her. Perhaps you'll remember that your job, right now, every minute, is to be her steward.

PART FOUR:

To Be a Witness Is Sublime

Your mind is the key thing here, the intricate machine that you alone have use of. It can hold the things you remember, allow you to love and cherish them, or it can drive you from the earth you knew and loved because you are afraid of what is coming.

We are many things; we are actors and angels, endowed with articulation and power, memory and empathy and cruelty. But, most remarkably, our minds are the instruments of witness. We've used them to see and know the earth, to chart her, study her, commune with her, and even to praise her. We've called our understanding "soul" and we've named the call to witness "god" and "blessing."

But, in the end, all those things are in us, in the mechanism that allows us to see, to perceive, and to know. We rest in the fluid place where we can watch marvels and thrill to the gentleness of the planet. We are perfectly designed to benefit from her bounty, to weep at her terror, and to ever see her as our mother, our redeemer, and our greatest work. We tremble before her, we work to subdue her, we worship at her beauty, and we poison and pervert her. We bend her and destroy her.

The truth is that all those things are of the same impulse, the impulse to prove that we were here, and that we knew her. That we were in dialogue with her majesty, that we existed.

And when we mourn the loss of the earth we've known, we are mourning the fact that we may not be here to perceive what will come next, to plumb the depths of what happens to this place, to record and grieve and rejoice, to see it all unfold. We are mourning the loss of our conversation with this place. We are mourning the witnessing we will not do.

It's been our blessing to be stewards here. As we have failed at that, so have we made a bargain to lose the right to witness and record all that we've seen. All those human minds playing their memories across the earth and farther back across all human time, they are the record of our responsibility, and the burden we have to act on what has been seen.

Really see a tree and you will want to tend it forever, to ensure the life it has staked a claim to. Perceive it as it truly is and you will weep for the grace that made it live and thrive, that made it whole and present in this moment. Really see it and you will wonder why you have been given the power to see.

And next you will think: I have been given the power to see because I have some important role. And maybe you will be right, and maybe not. Maybe the presence of that perception is its own reward and we have been right to take it and give little back.

But, maybe not. Maybe holding these perceptions has been our sacred trust. Maybe the reason that we exist at all is to house those perceptions. After all, they have become the spiritual meat of the planet. They make it special. And without us to know the place, love the view, chart the change, worship the sublime, the earth would have lost a great deal.

But we could make ourselves relevant to the earth, live up to our gifts finally, if we take that role, that knowledge and trust and truly live it. We could find our purpose as the caretakers of this place. We could truly live the purpose of our perceptions.

To bear death, we must be alive to life. To bear the ruin of the earth, we must be alive to its glory. To know its glory is the only way to find the strength to fight. And, in fully perceiving what we love, there is a path to save ourselves.

PART FIVE:

An Elegy for What Came Before

Here's how I remember what I've loved. I start with an image. An image can live in the mind, even bifurcated by multiple viewings in memory, until it becomes an icon, a talisman, a ritual.

That close-up of the blueberry bush I remember from when I was four, where I stood and ate until someone came and found me. The light in its leaves and branches—just that—could be a little filmstrip of how the color and wind looked while the seconds ticked by.

Multiple rememberings only stretch the movie, fade sections of it, feed the image back to itself. Like a loop, like a mantra, like a prayer, it tells me I existed once, and that I saw the earth. I was of it, standing in a moment of its grace. You have this too, images of the earth as it was once, where, less abused, it thrived.

Think how many of these images there have been—in private stockpiles, throughout human memory. Once there was someone living on the earth who had seen the empty Kansas plains, the island of Manhattan covered with forests, the rich dark woods of Germany, full of wolves. Once, for comfort, someone played a picture in their mind of pacing through a swampland full of turtles, a meadow where a moose stood, a day when you couldn't hear an airplane in any sky.

Now we are those who will someday say, yes, I saw the coral reefs, the island chains of the Pacific, the glaciers of Argentina, the coastline of the Gulf of Mexico, Bangladesh.

Memorizing them brings both a pleasure and a pain. To be with the land, the cities, and the people as they are, you have to be present, with them, immersed in cloud or ice or water. But the preemptive nostalgia of the knowledge of their loss will make this tricky. And the dissonance of elegy and experience existing in the same moment may make the work of holding these memories seem steep and monstrous.

The time before, after all, was a beauty of variation within the predictable.

I try to remember what I've loved. Birches marched across the edge of a hill, black rocks collected the foam of active wave, the ocean calmed and let its ripples lengthen, clouds stacked under airplanes and sent out mist.

I take myself back to the arching dunes above the Atlantic, beach houses and sedge grass along the placid ocean. I stand again on the Mer de Glace. Its bulk filled up the valley with dirty, ancient ice, and I was there to see it as it moved but didn't move. I remember.

Images stack up and multiply: random commerce, the bulk of some gorgeous city with power pouring out of windows and signs, the snowfalls of my childhood, the gentle weather and the regularities of seasons in a Massachusetts forest. There are too many things to count, and I am steward to them all.

Now you try. Make a filmstrip for yourself—a Joshua tree arches fingers at the sky, a set of stones click together in a springtime stream. Add to them every other benign and stunning image of the earth that you can find and all the casual moments of culture. Stretch them like an unbroken movie of the way you used to think.

Now notice that the movie fractures, you can no longer see that image as you once did. Now the interpretations of the time after blur and break and hiccup the image until you see it as a different thing.

Hold it, replay it, see the change and the way it works on you. This is what you have become the steward of.

echo

The planet must echo

with our ghosts, ghosts and

with all the things we did a

and wanted still to do.

to do.

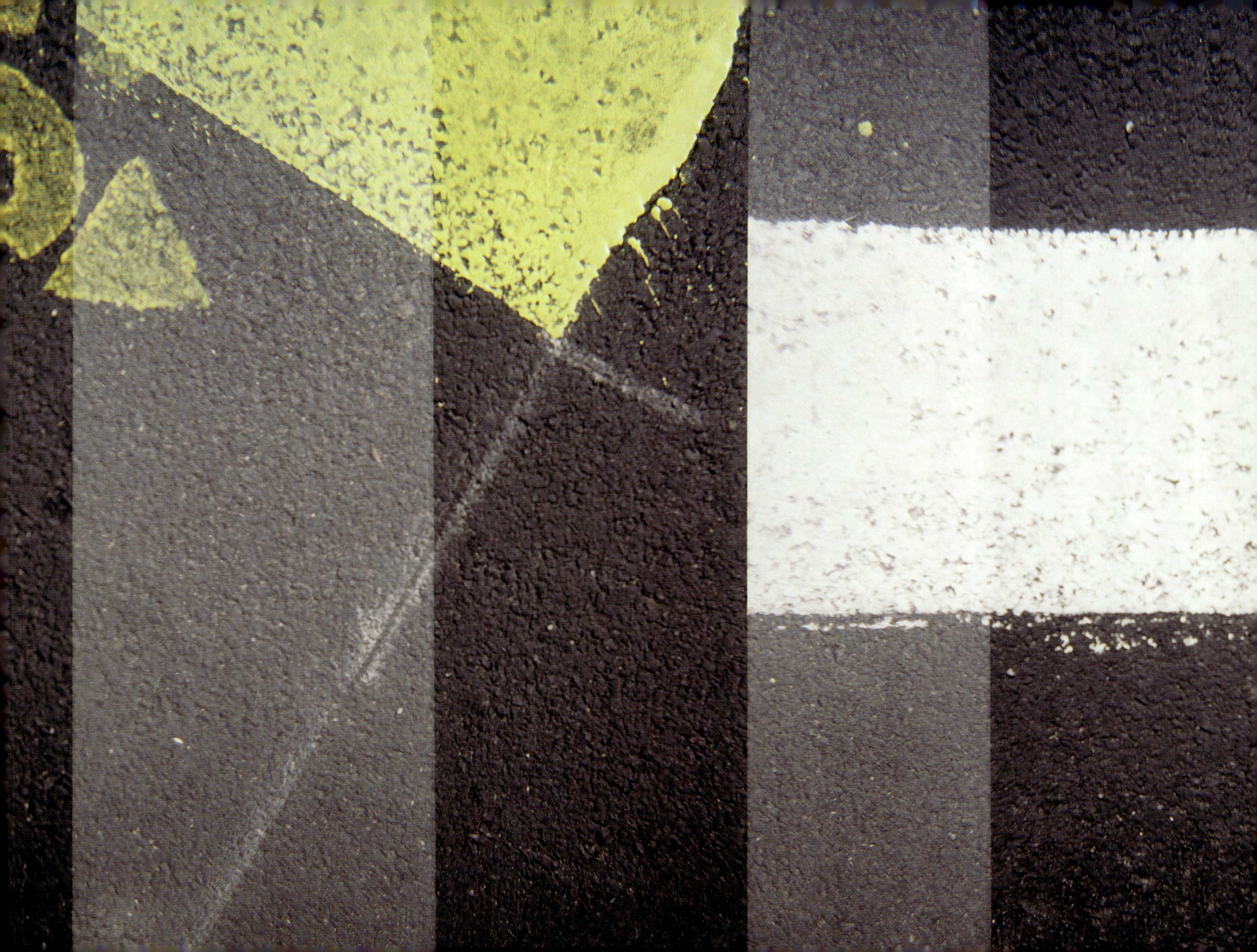

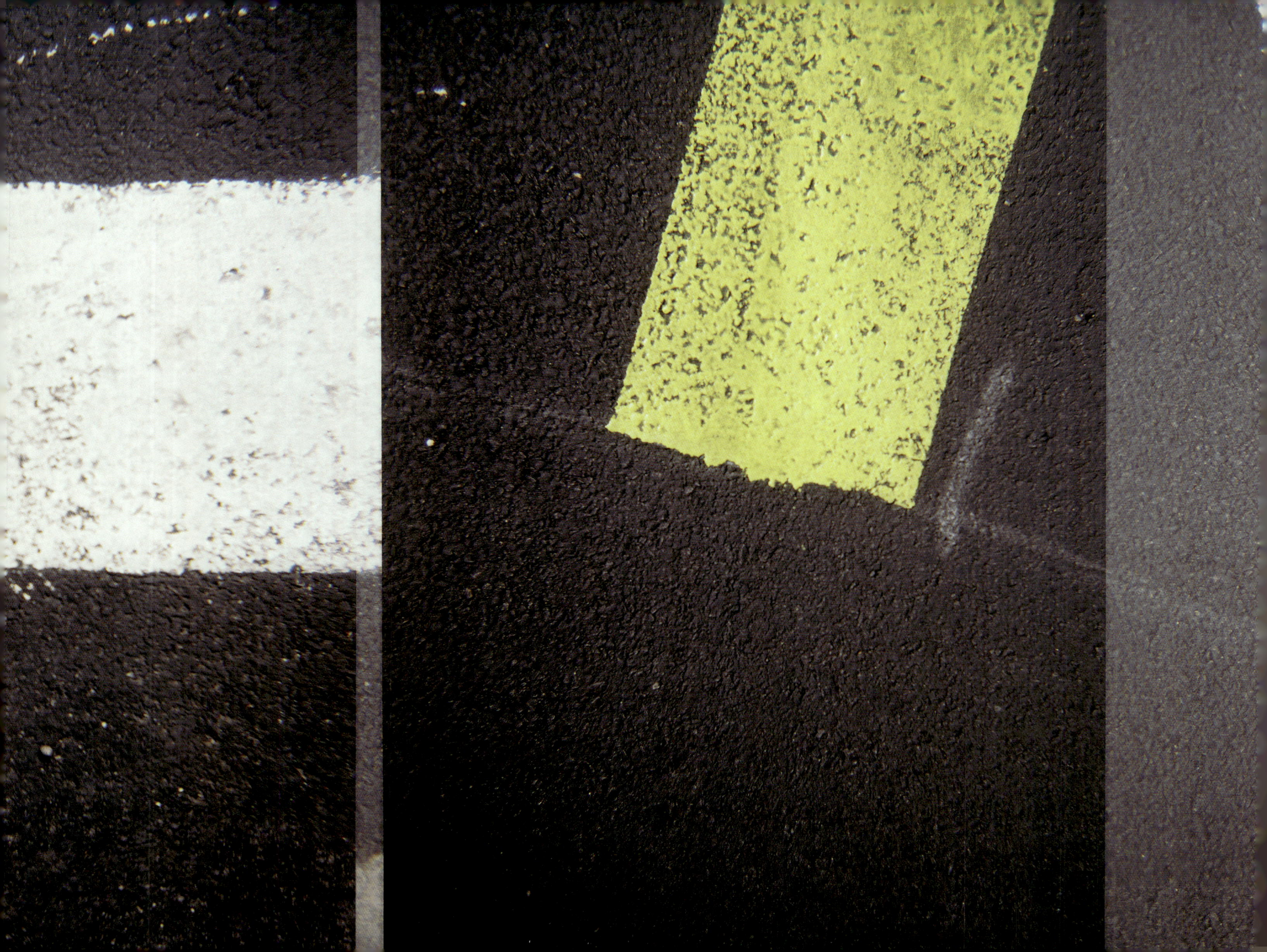

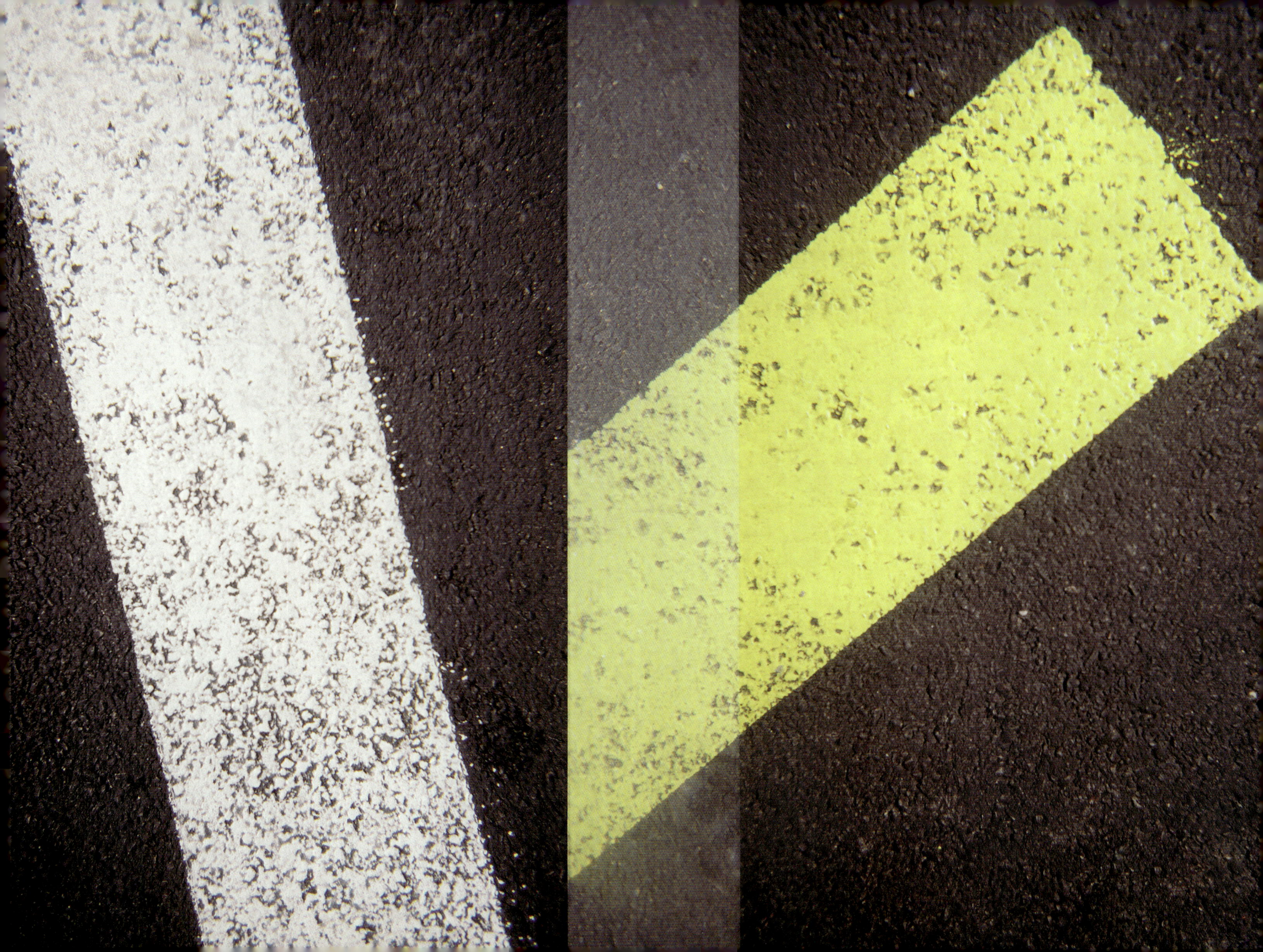

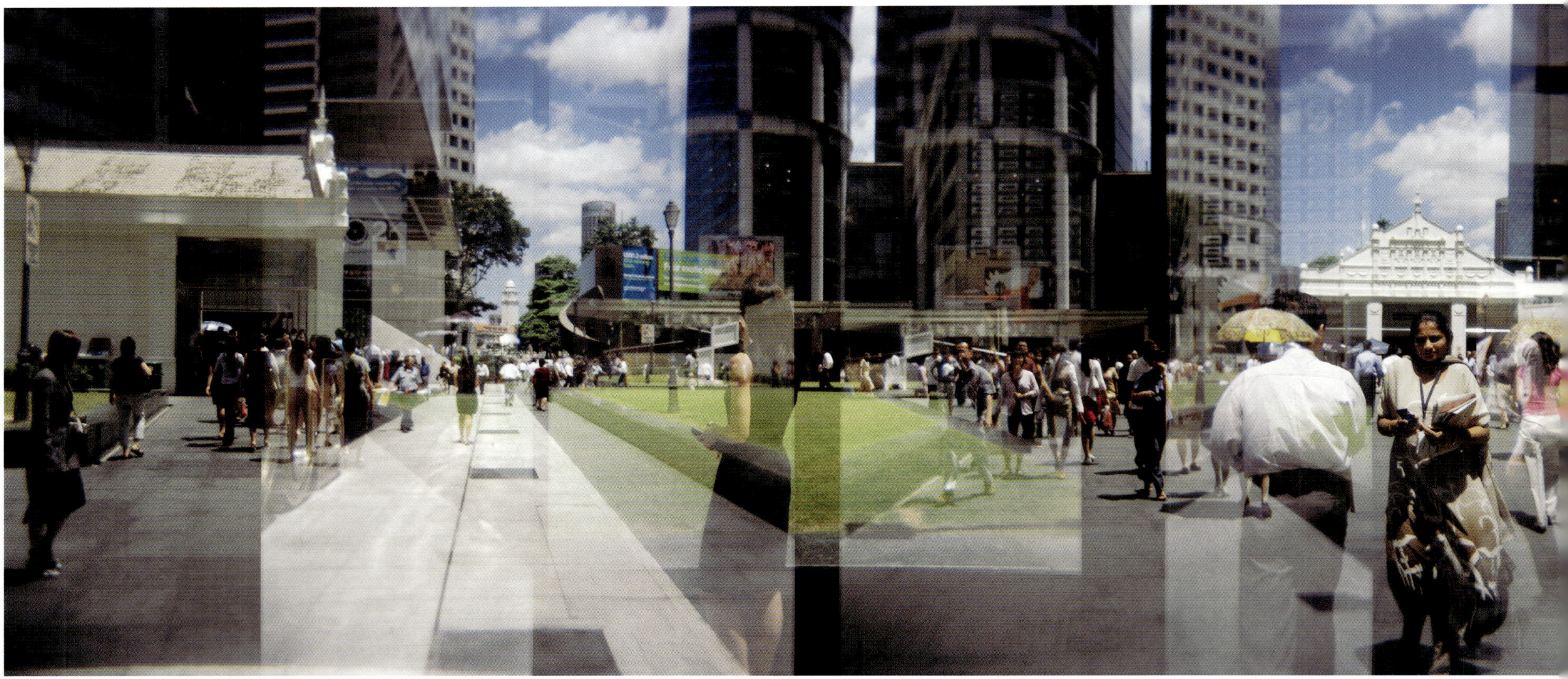

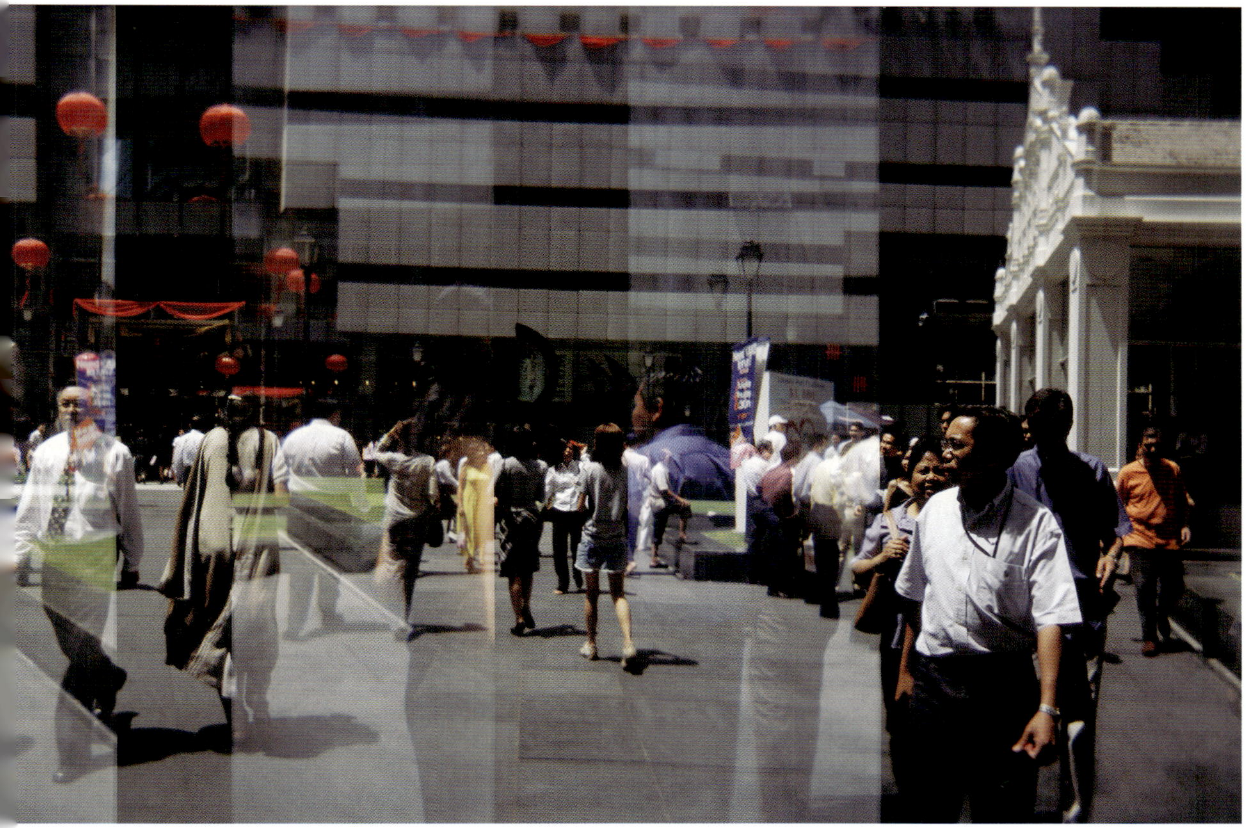

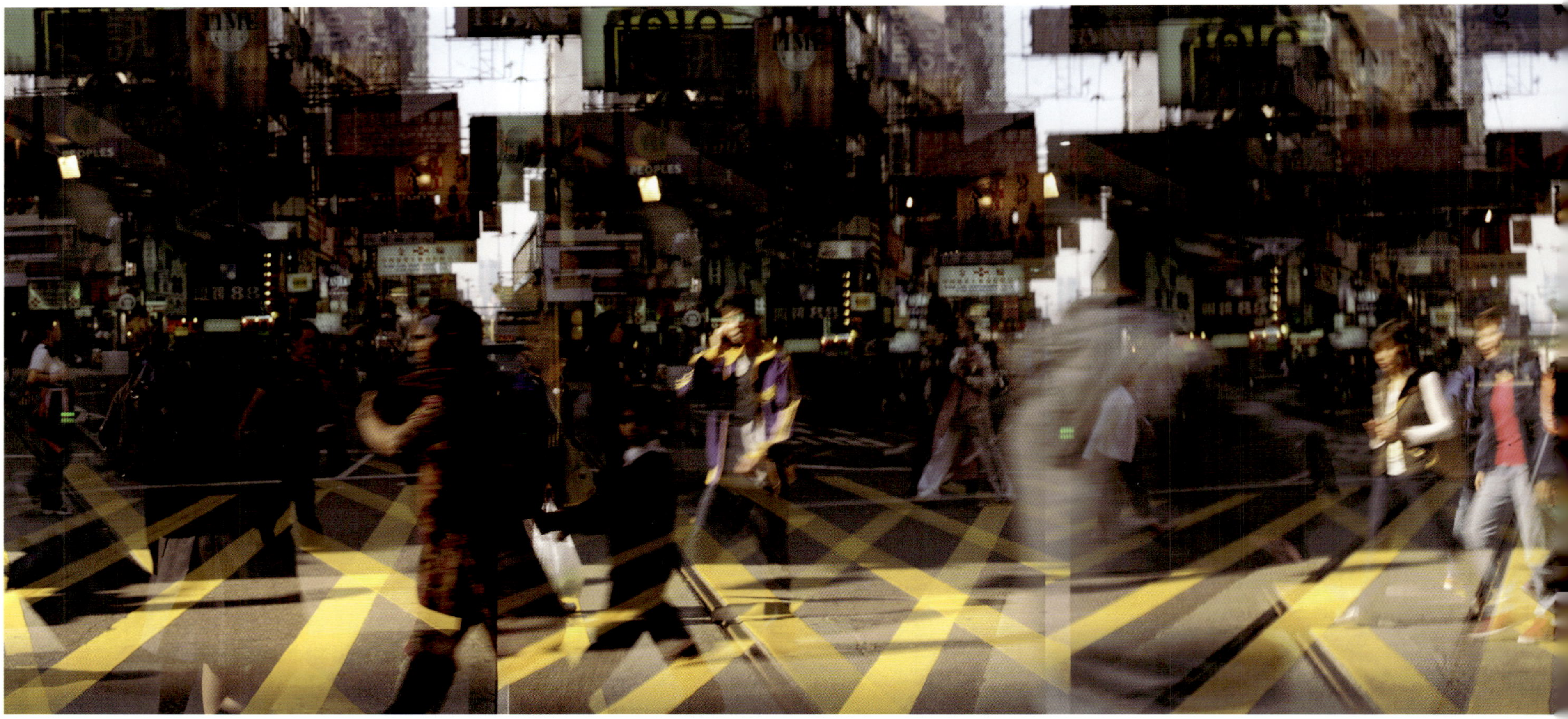

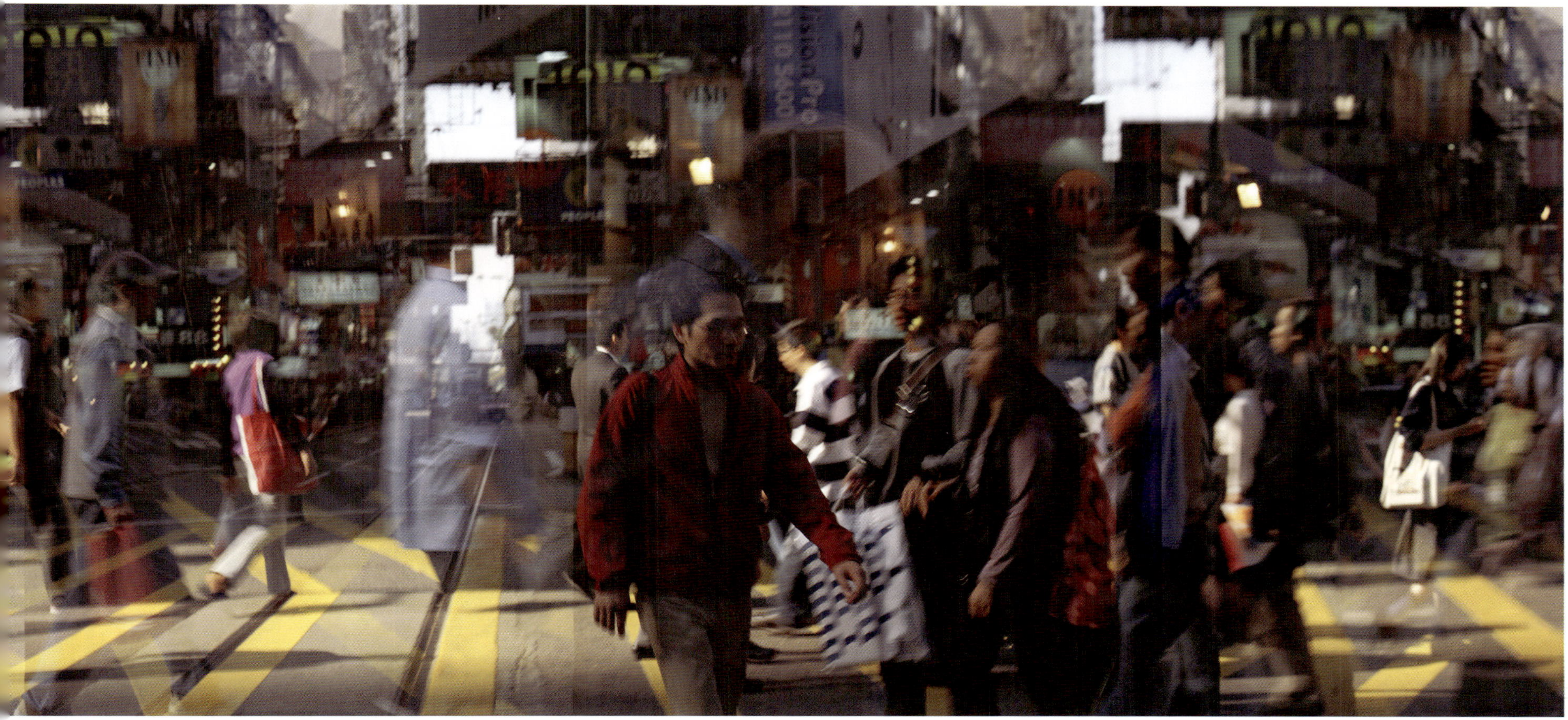

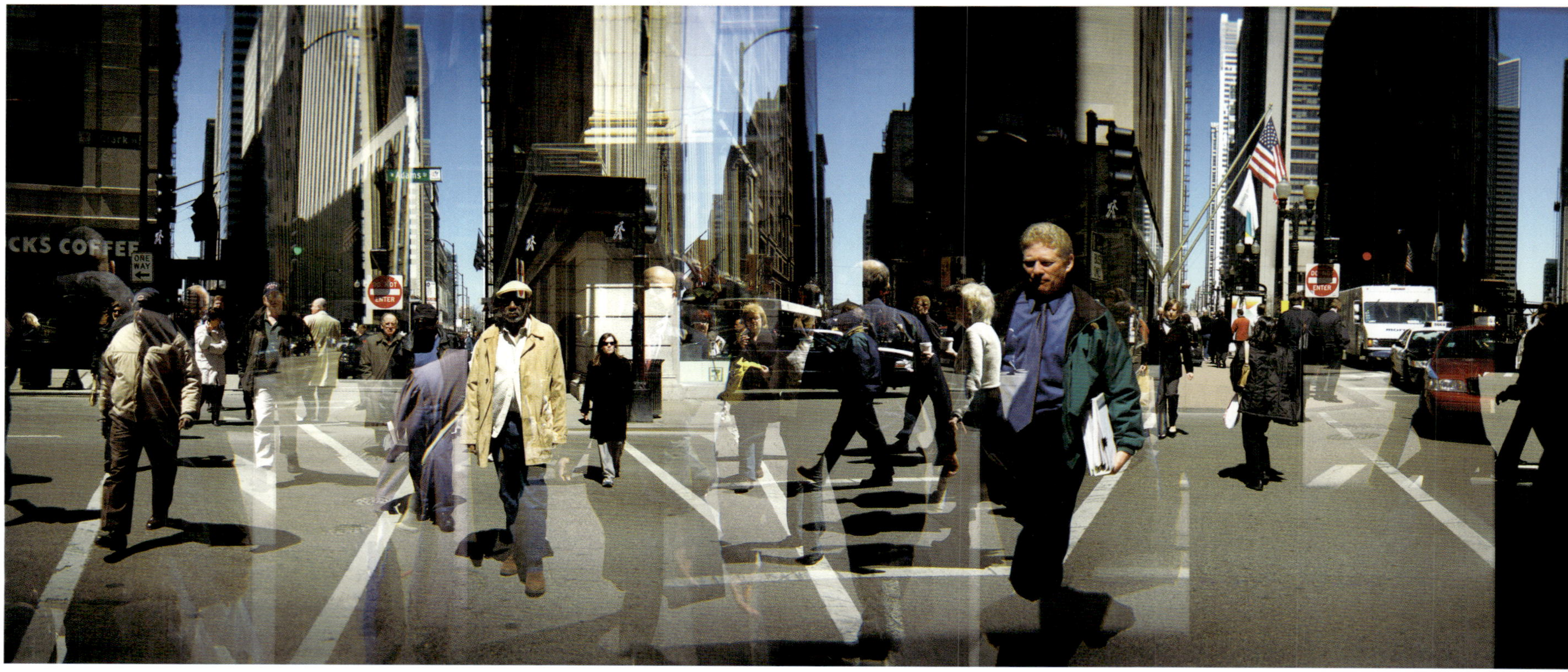
ONE WAY
DO NOT ENTER

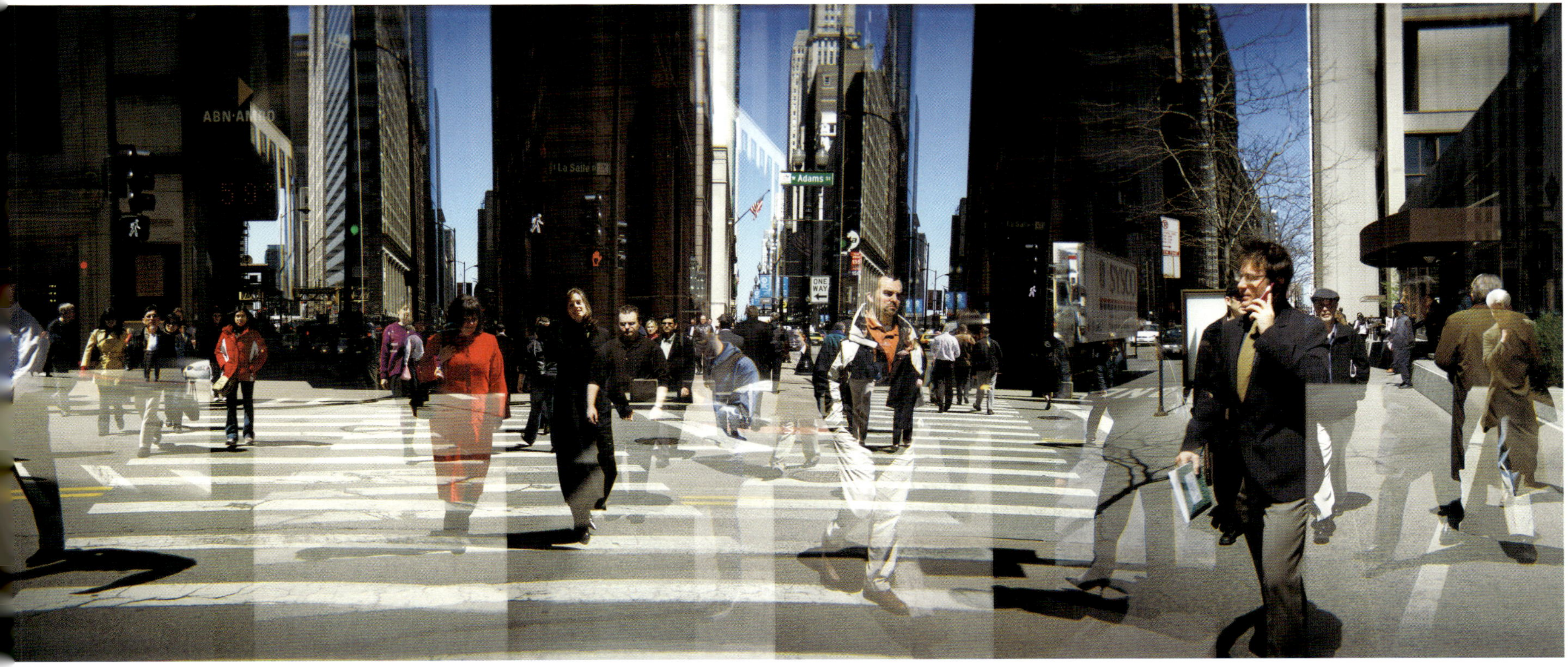

ABN·AMRO
Adams St
ONE WAY

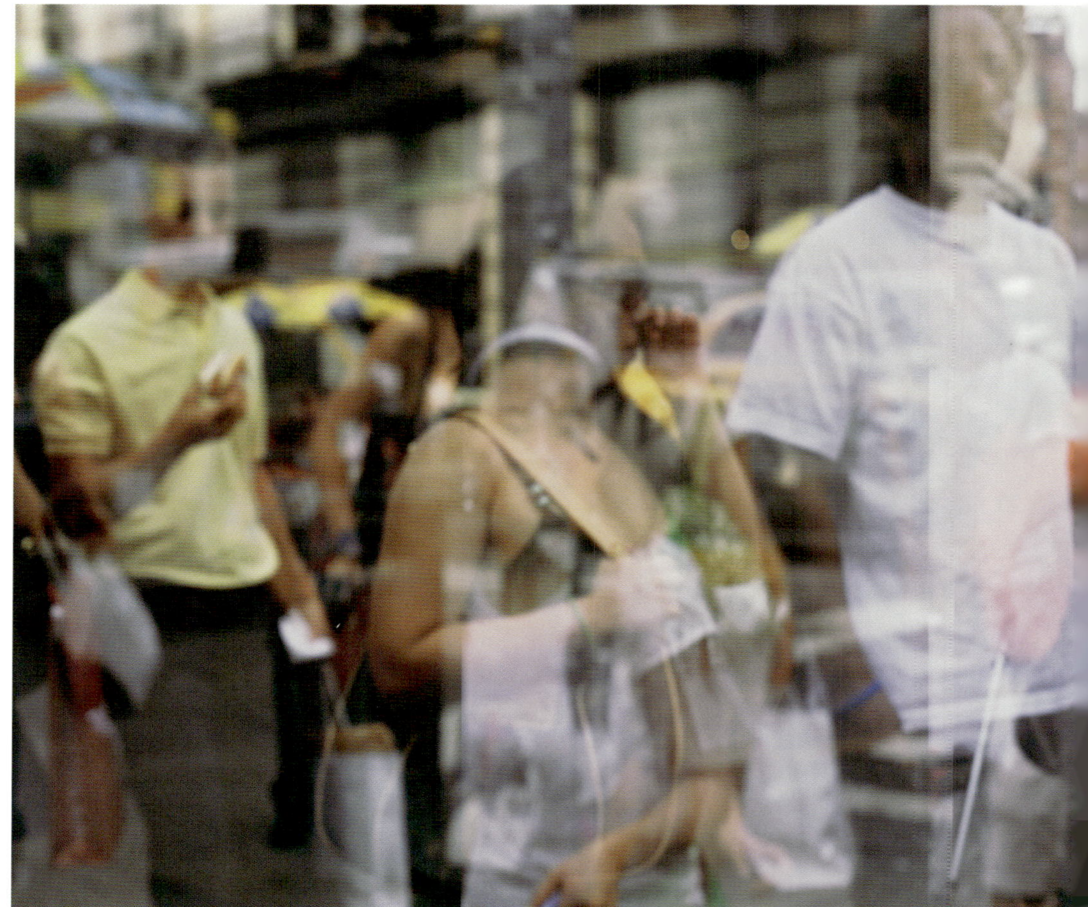

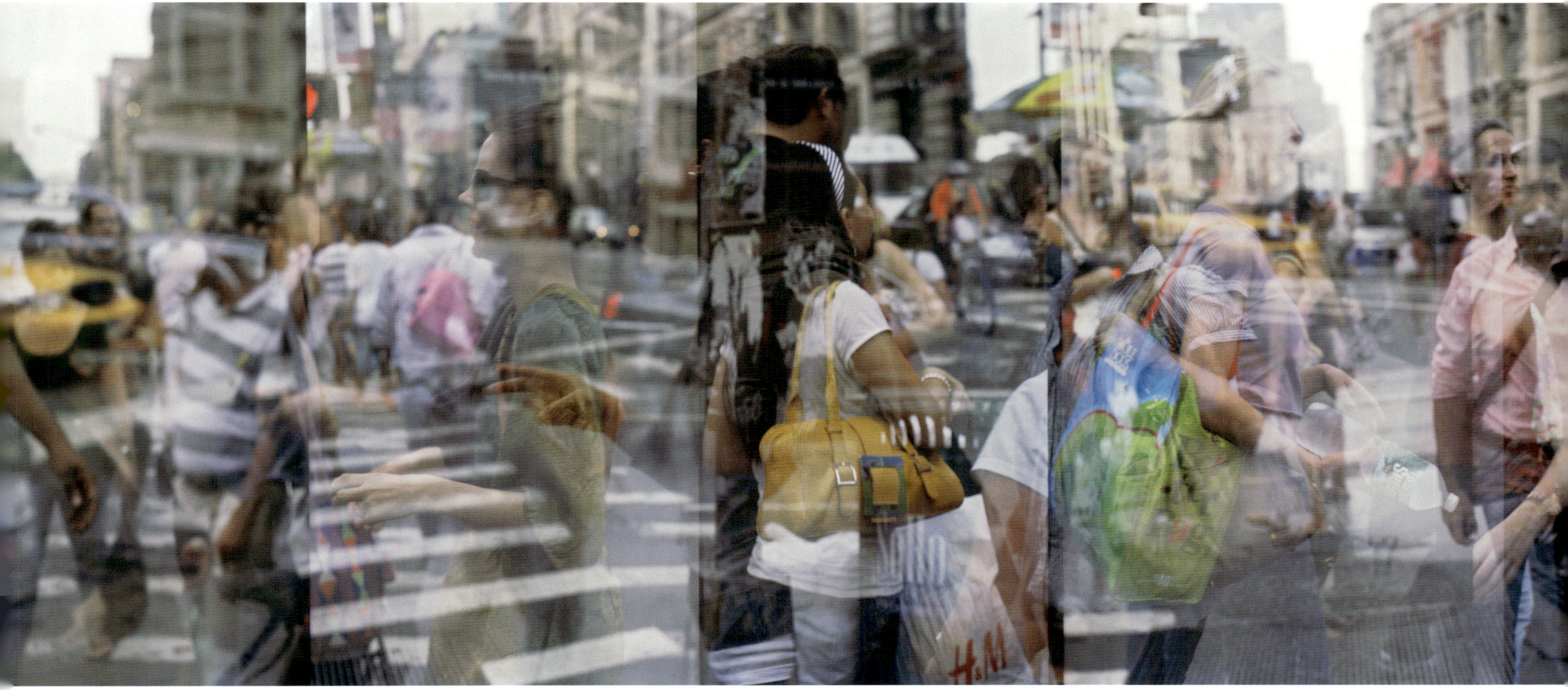
H&M

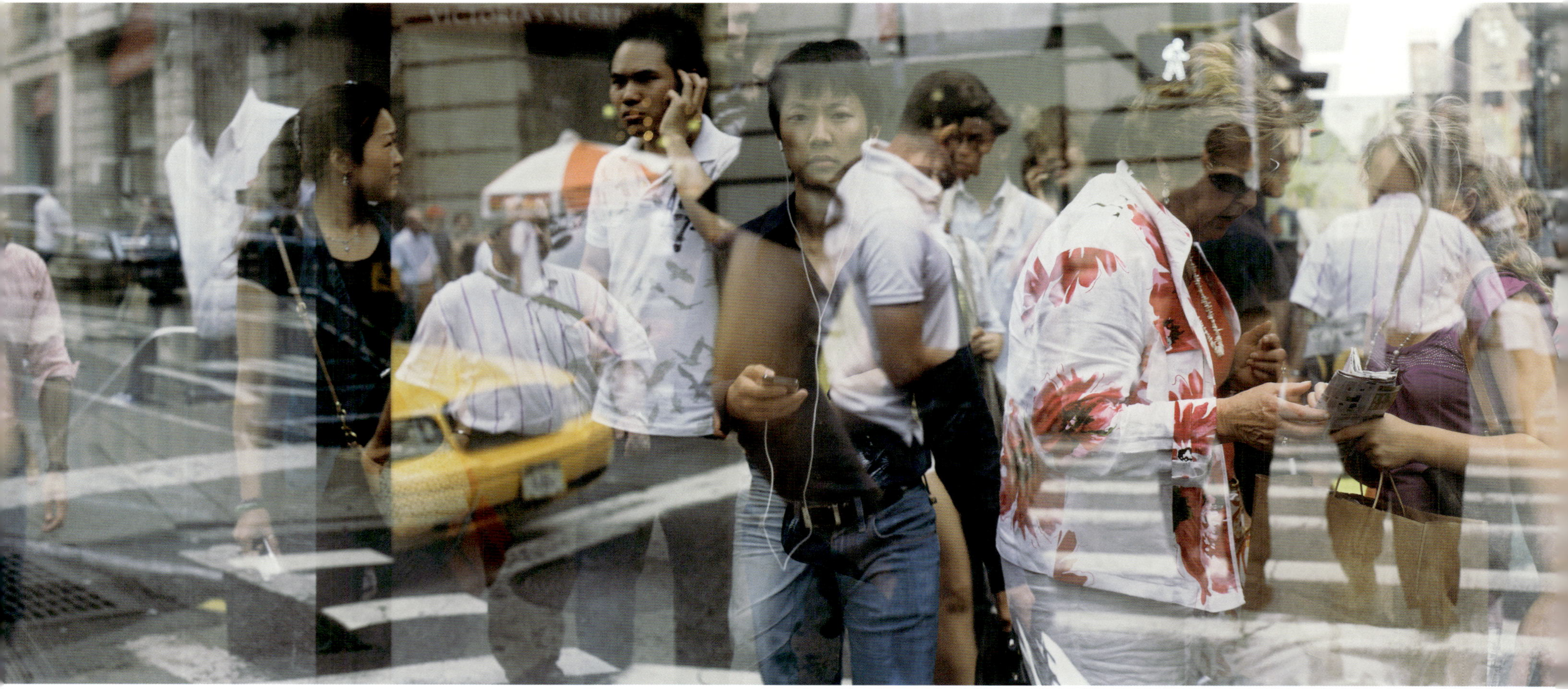

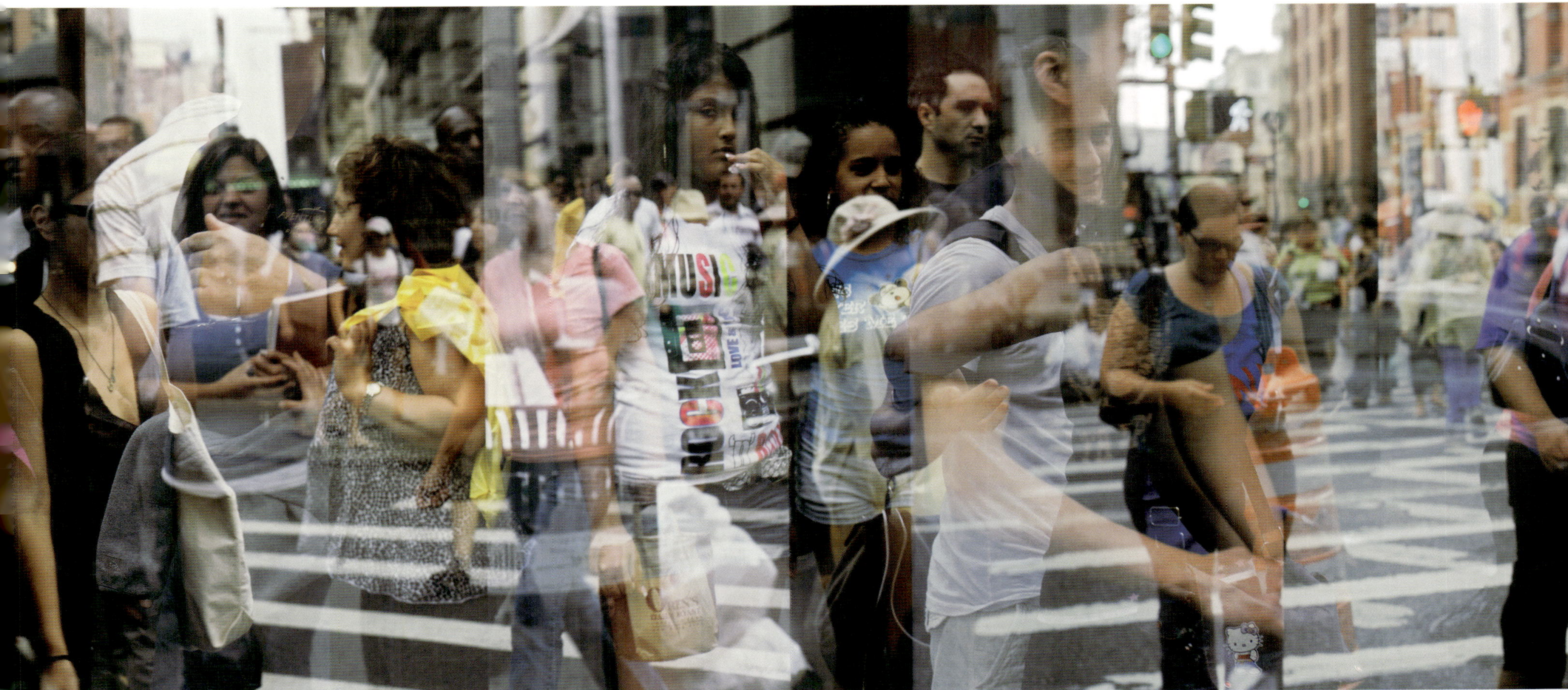
MUSIC

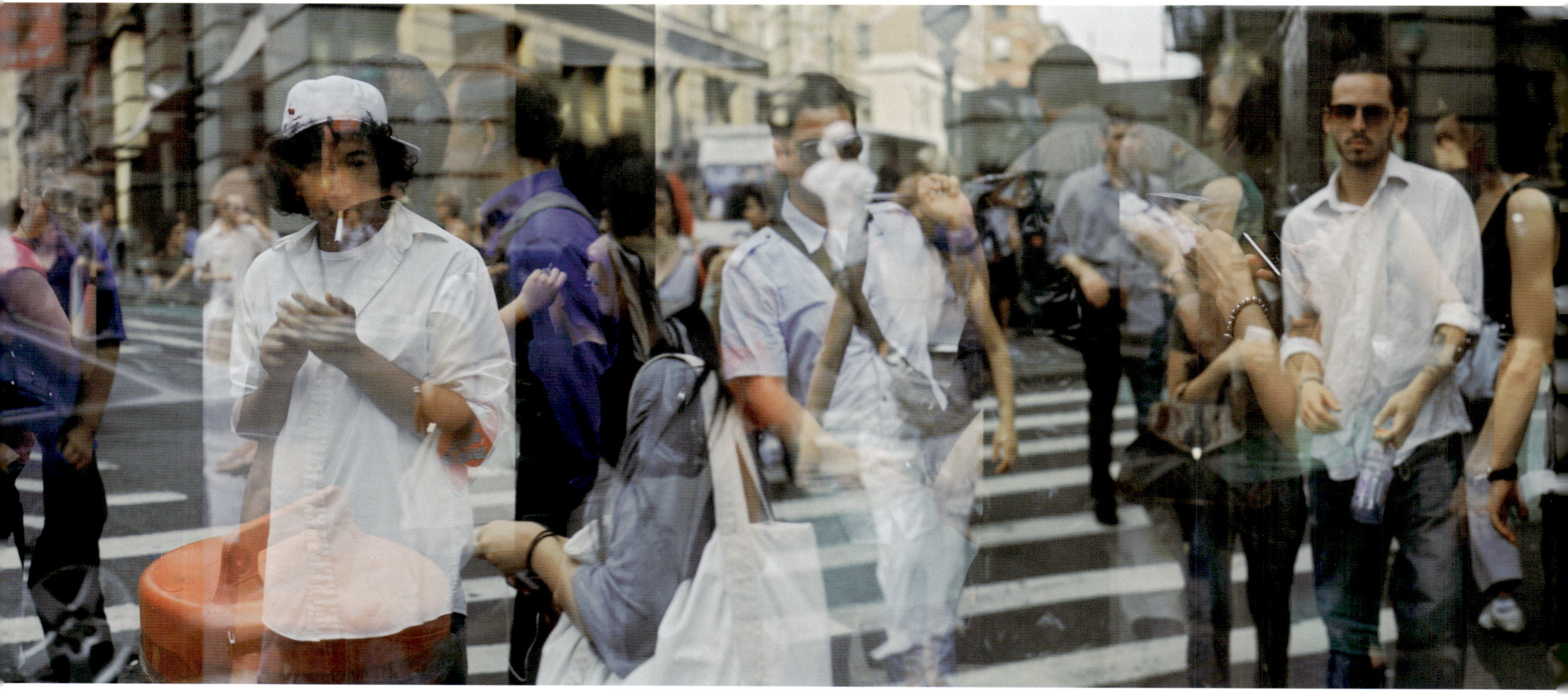

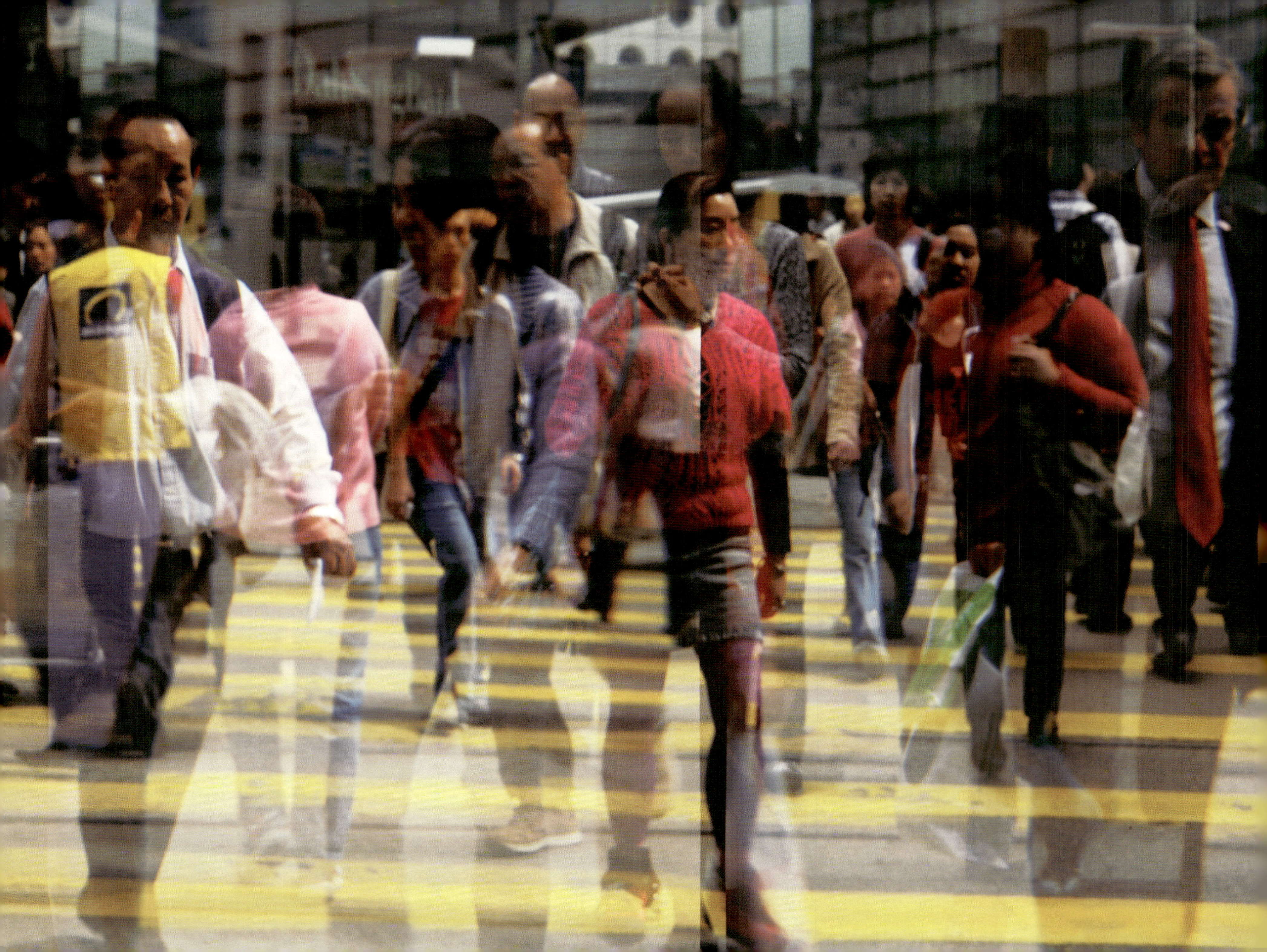

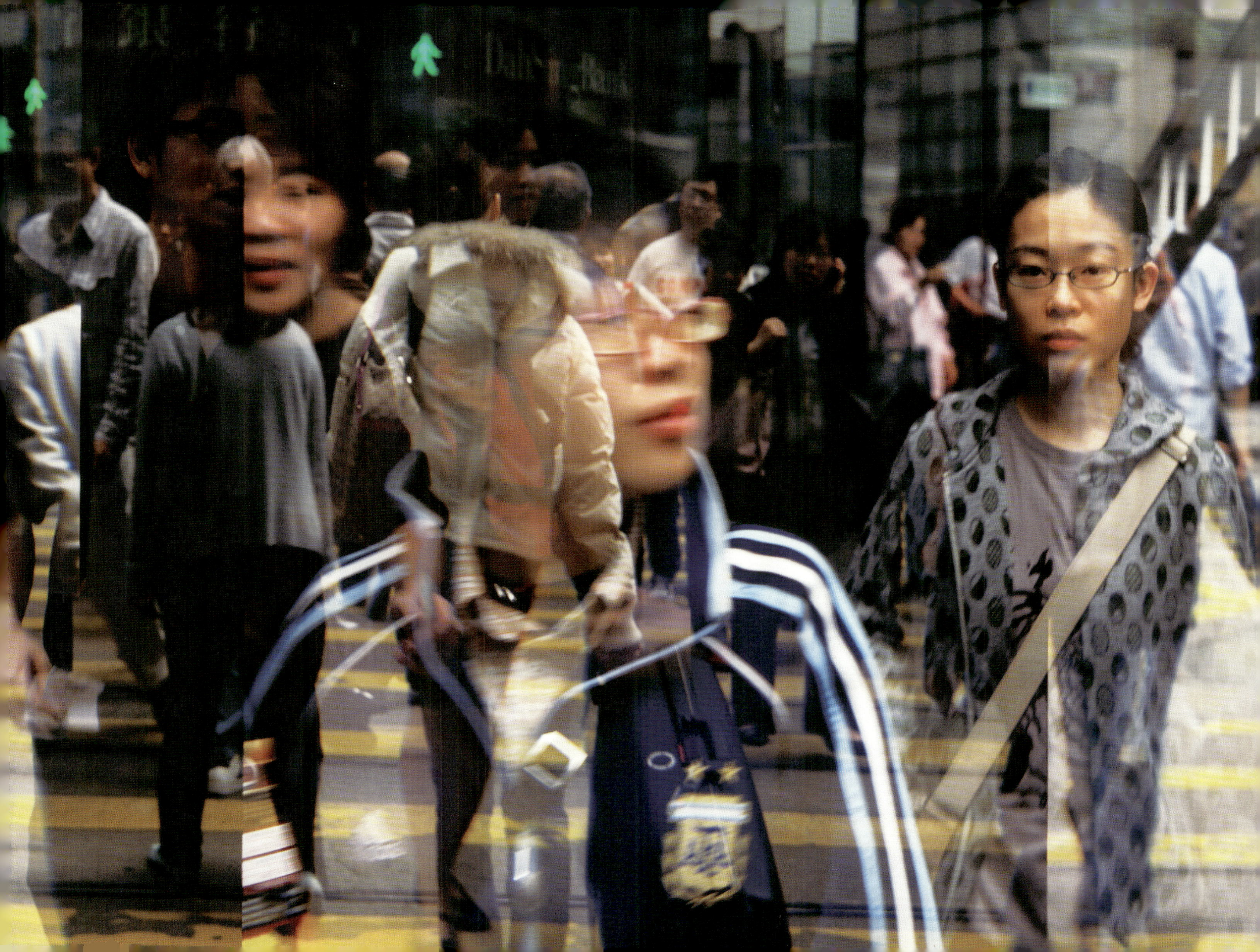

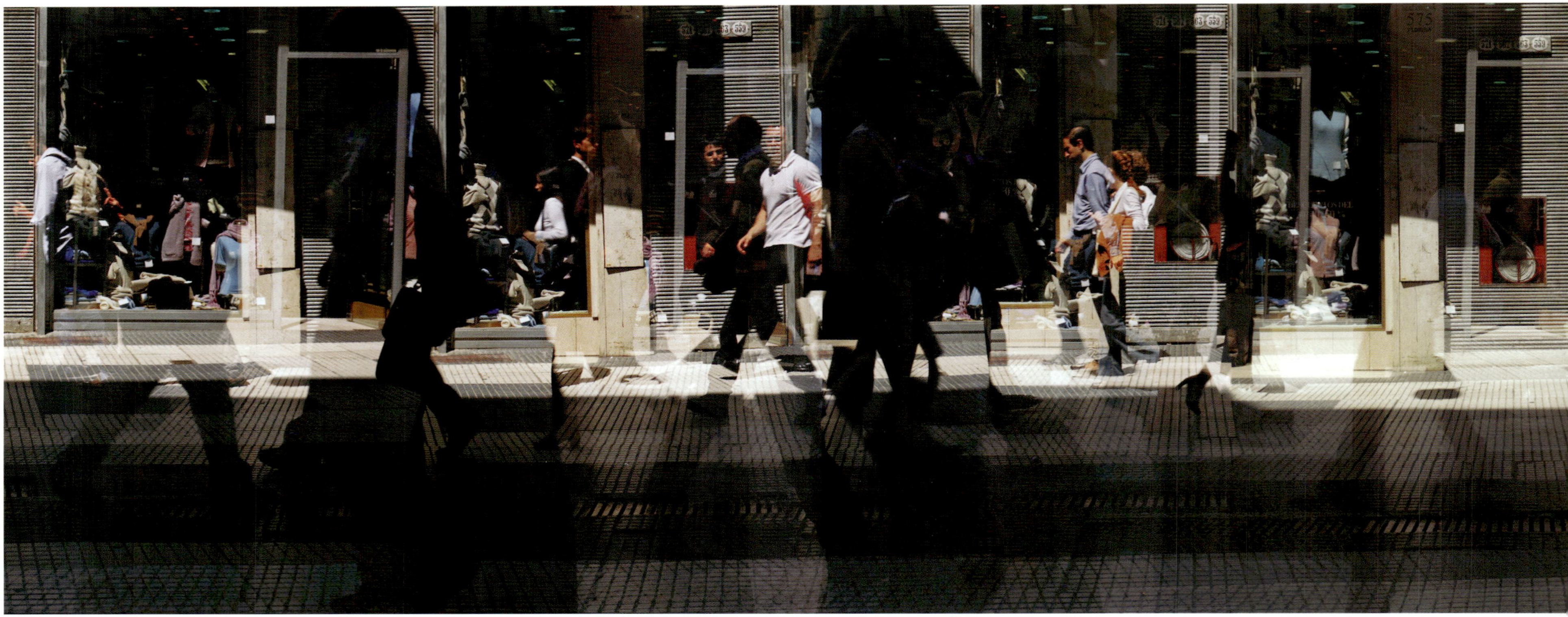

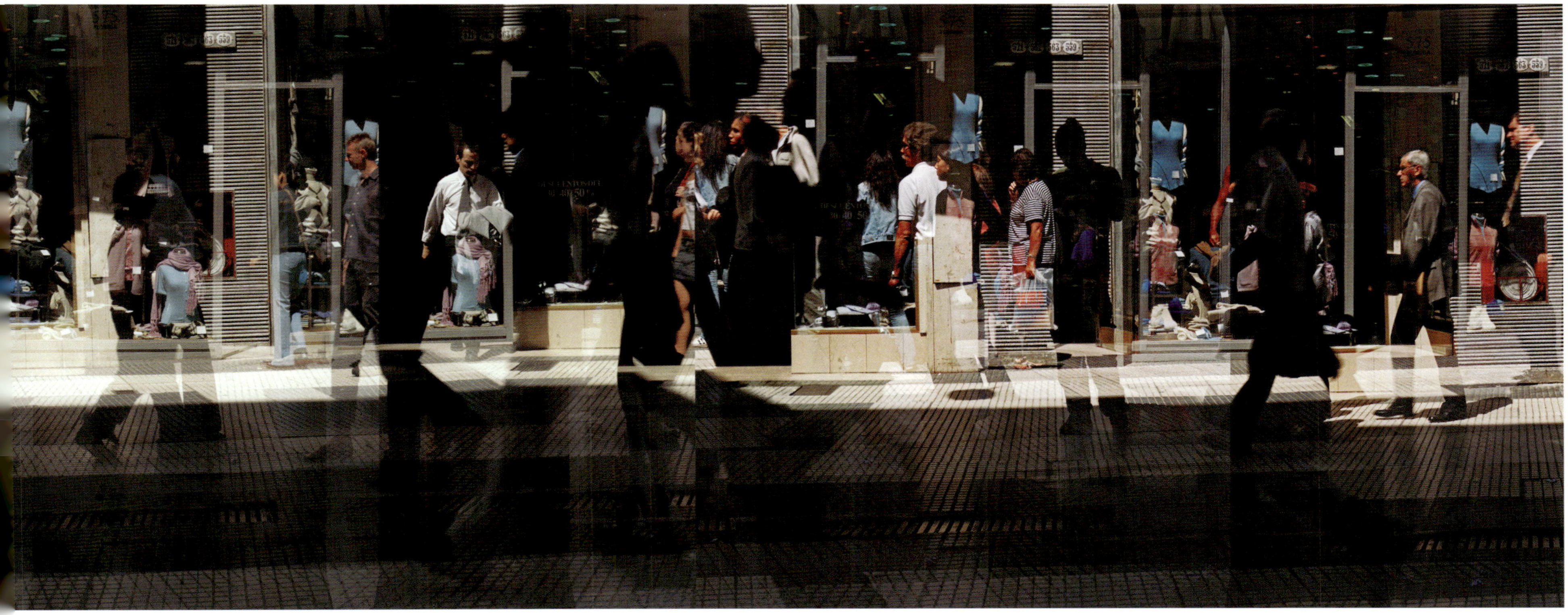

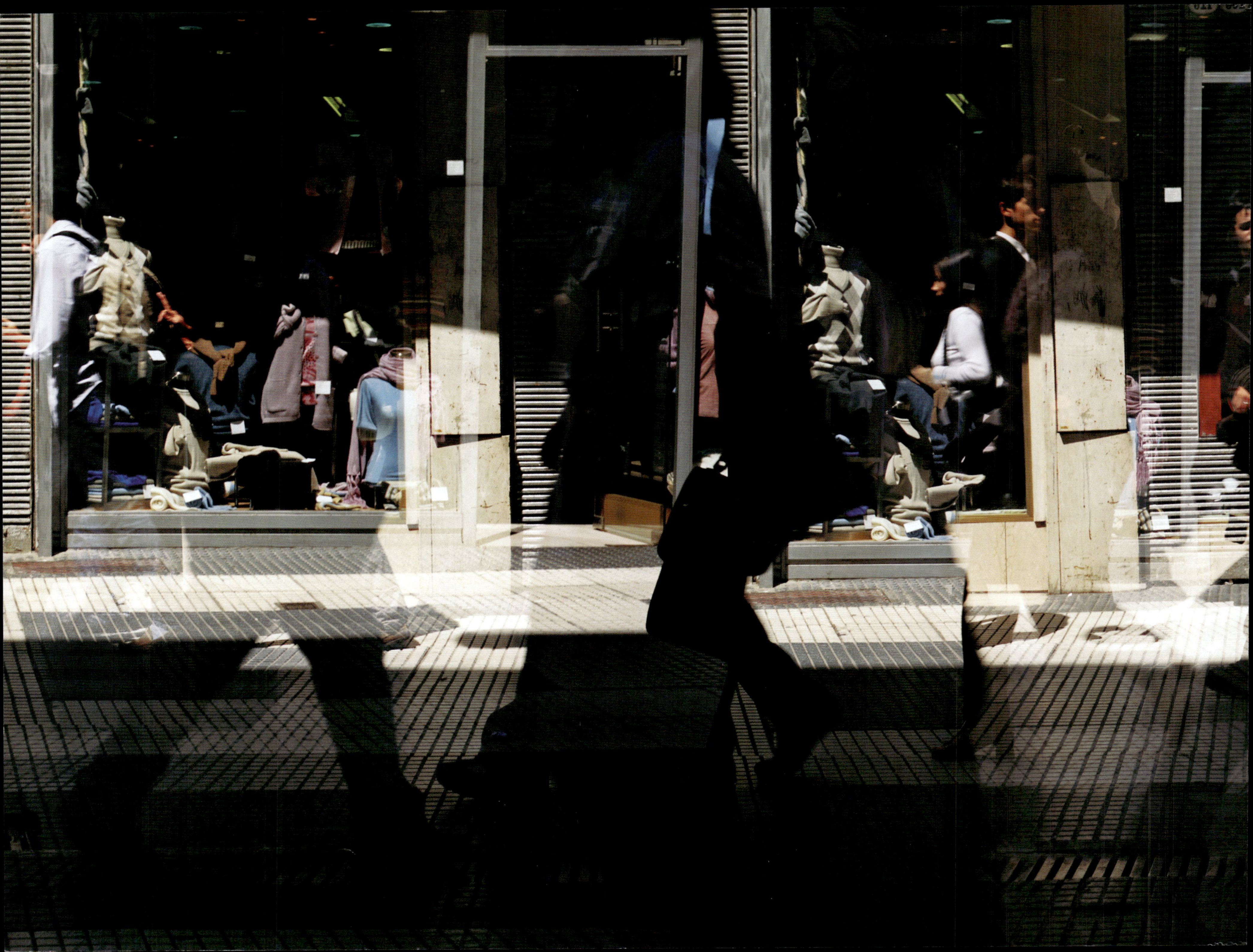

IN THE TIME AFTER.

by Derrick Jensen

In the time after, when industrial civilization is a bitter and too-slowly-fading memory, a memory of a nightmare too atrocious to be believed by those who were not alive in the time before and so did not experience it and its destructiveness, birds will begin to come back, and whip-poor-wills will sing, and bobwhites will sing, and murrelets will fly to oceans no longer being murdered and will return with their bellies full of fish to feed their young. In the time after, frogs will sing and nights will be as full of their songs as it is possible to be. Nights will be even more full than that, until the nights burst from their fullness.

Early in the time after, many of those who survive will barely believe their fate. There will, of course, be those—a small minority (they were always a small minority, even before the fall of civilization, no matter what they pretended, no matter what they made believe)—who look longingly back to before and wish they were still dead; who wish they were still killing all others so they could cling to the comforts or elegancies that made life, to their way of thinking and not-thinking, worth living, that made life bearable, even if sometimes only barely. But they are, truly, an insignificant minority.

For many of the others, for the overwhelming majority, the first feelings were perhaps a shocked and disbelieving relief, the emotional collapse that so often comes after a terrible tragedy is narrowly averted and you no longer have to be quite so strong—a sudden crumbling of the prison where you waited for the unjust death sentence to be imposed.

After that came a cautious optimism: the relief might be real, the threat gone. Then later, when the enormity of what had happened—and even more, what could have happened but did not—finds its way fully into their bodies, these others will feel a joy deep and profound enough to soften at least a little of their sorrow for those who were lost, those who did not make it: passenger pigeon, great auk, dodo, shoal sprite, chestnut ermine moth, Round Island burrowing boa, golden toad, Snake River sucker, Central Valley grasshopper, Falklands Islands wolf, quagga, Lake Pedder earthworm, Cape Verde giant skink, silver trout, Galapagos amaranth.

In the time after, cars will no longer hurtle down roads—free-fire zones for these massive metal projectiles—killing everything in their path. Machines will no longer hum. Chainsaws will no longer whine. Humans will eventually no longer value pseudo creation—production—more than they value real creation—real life, their own and others'. In the time long after, humans will no longer sell or buy time, their own or others'. They will no longer sell or buy any of the other things—beings—who cannot truly be sold or bought, things—beings—like land, like trees, like fish or frogs or birds, like other humans, like ideas, like art. In the time after, life will be more important than machines, production, commerce, power.

In the time after, wounds will scab over, begin to heal. Dams will fail and rivers will again run free, will again live. In the time after, skyscrapers will fall in on themselves. Roots will buckle concrete and tear up asphalt. Forests will begin to expand, to reclaim what was once and forever theirs. In the time after, flocks of birds will grow, not dwindle. Schools of fish will grow, not dwindle. The same is true for herds of bison, packs of wolves, communities of prairie dogs, runs of salmon, lamprey, grayling.

In the time after, humans will remember how to be human, and will again listen to the voices of those we, the civilized, have so long silenced. Humans will again hear the voices of stars, soil, rocks, fire, ice, rivers, seas, forests, grasslands, deserts, and so many others, and all who live in them. Humans will, individually and collectively, give back more than they—we—receive.

In the time after, there will be no more time as we now understand it (or rather, as we make believe we understand it). There will be days and nights, and springs and summers, falls and winters. There will be markings of seasons: the singing of frogs, the blooming of trillium, the return of geese, the play fighting of baby bears who use any startling excuse to scamper up trees, the flowering and fruiting of so many plants, the yellowing of leaves, the return of rains, the return of salmon, the fruiting of fungi, and the singing of frogs to begin it all again.

In the time after, there will be all of these and so many more that we could not possibly know them all, no matter how closely we paid attention, no matter how long we lived in and with a place, no matter how well we listened and learned. There will be all of this, but there will be no time, no ticktock of clocks, no artificial constraints, not only on our time, but on time itself. In the time after, reality will determine time, and not the other way around.

But it will take a long time to return to this time of no time: the people during and immediately after will still be as they are now, which means they will still attempt to silence all others, will still exploit and consume all others. They will still attempt to enslave all others that they perceive as below themselves, and will still enslave themselves to those they perceive as above them, who will in turn enslave themselves to those they perceive as higher still, and so on, with all of them enslaving themselves most of all to the whole system of enslavement. And in the time after, the humans still living will still allow themselves to be enslaved to separation.

After a time—after a very long time—these humans (or rather their many-greats-grandchildren)—will unlearn their relentless need to be enslaved and to enslave, and they will also unlearn their enslavement to their notion of time. They will even unlearn their enslavement to separation.

In that distant time after, people will no longer believe they are separate. Nor, of course, will they believe they are one. Both beliefs will be perceived as being as unrealistic as in all reality they are. If you and I were both alive in that distant time after, you would still be you, and I would still be me (as we are now). But we would recognize that we are neither separate, nor are we one. We would perceive that we are as we are: permeable.

And we would perceive that time, too, is permeable, porous; that time is no more linear than a river, that it rushes, slows, eddies, twists, floods, pools, soaks into soil, and permeates air; that time is complex, like a forest, like a prairie, like a desert, like a slough or lake or ocean trench; that it doesn't progress so much as it grows, becomes tangled; that it flowers and bears fruit, and dies back and regenerates. Time. Not just the effects of time. But time itself.

People—of course, including nonhuman people—will understand all this in the time after.

In the time after, time will not be something to be measured and controlled, confined and subordinated, something to enslave and at the same time be enslaved to. Instead, time will be a cave to be entered and explored (or not); or a lake into which you dive and swim (or not). Time will be food you eat, water you drink, air you breathe. Time will be where you live and who you are. Or not.

In the time after, time will be what it always has been—no matter what we have so frantically made believe—which is life, as solid as the flesh of rocks, as gracious as the flesh of clouds, as graceful as the spaces between ocean waves. In the time after, we will remember how and why to fall back into these spaces, and we will remember grace and graciousness and solidity, and once again and at long last we will realize that we are no longer and have never been separate, and once again and at long last we will rejoin the living.

That will all come, and will all be, in the time after.

remember

humans will remember

how to be human,

and will again again listen to th

ne voices of those we the civilized

have so long

e the civilized

have so long

silenced.

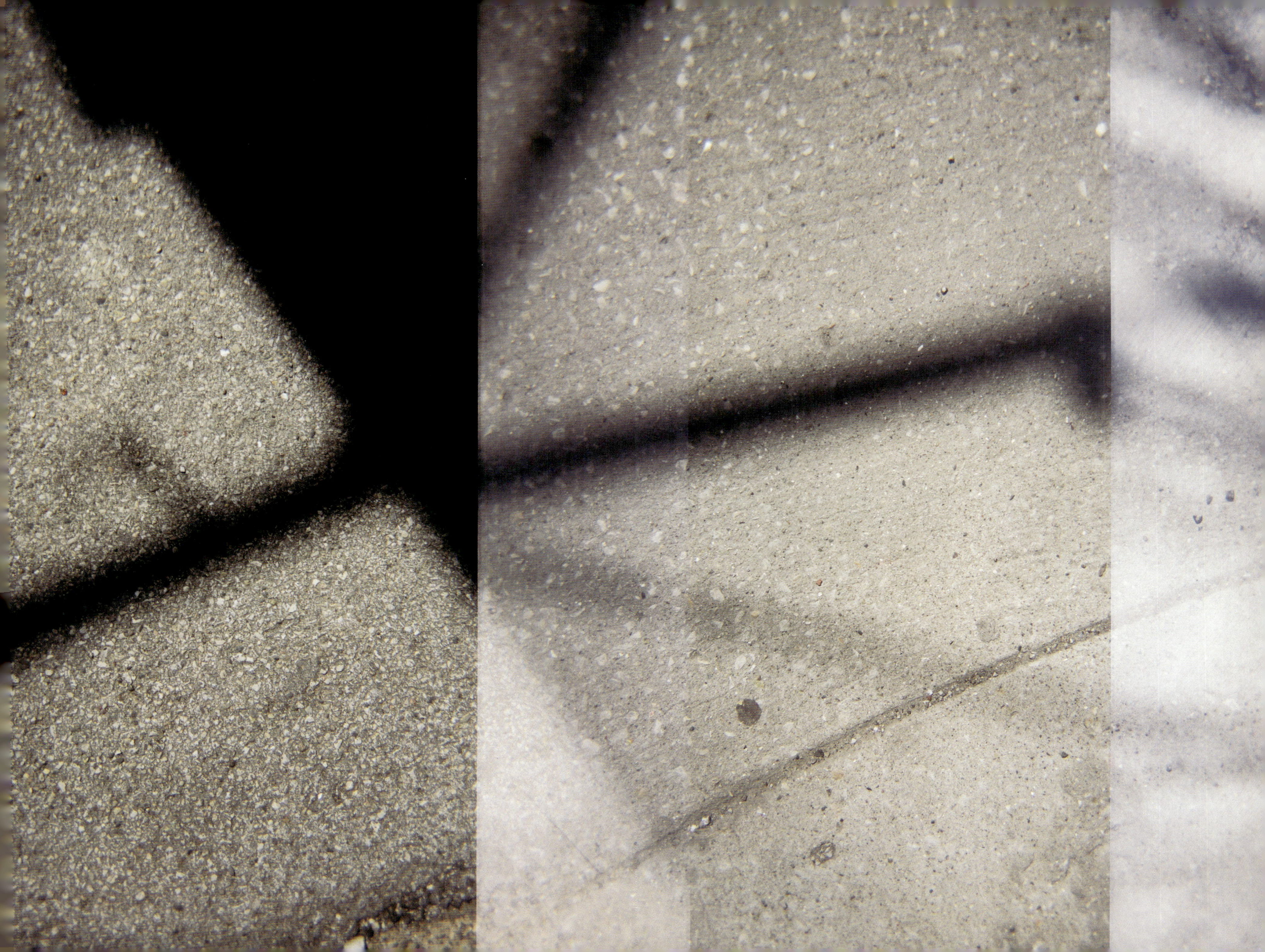

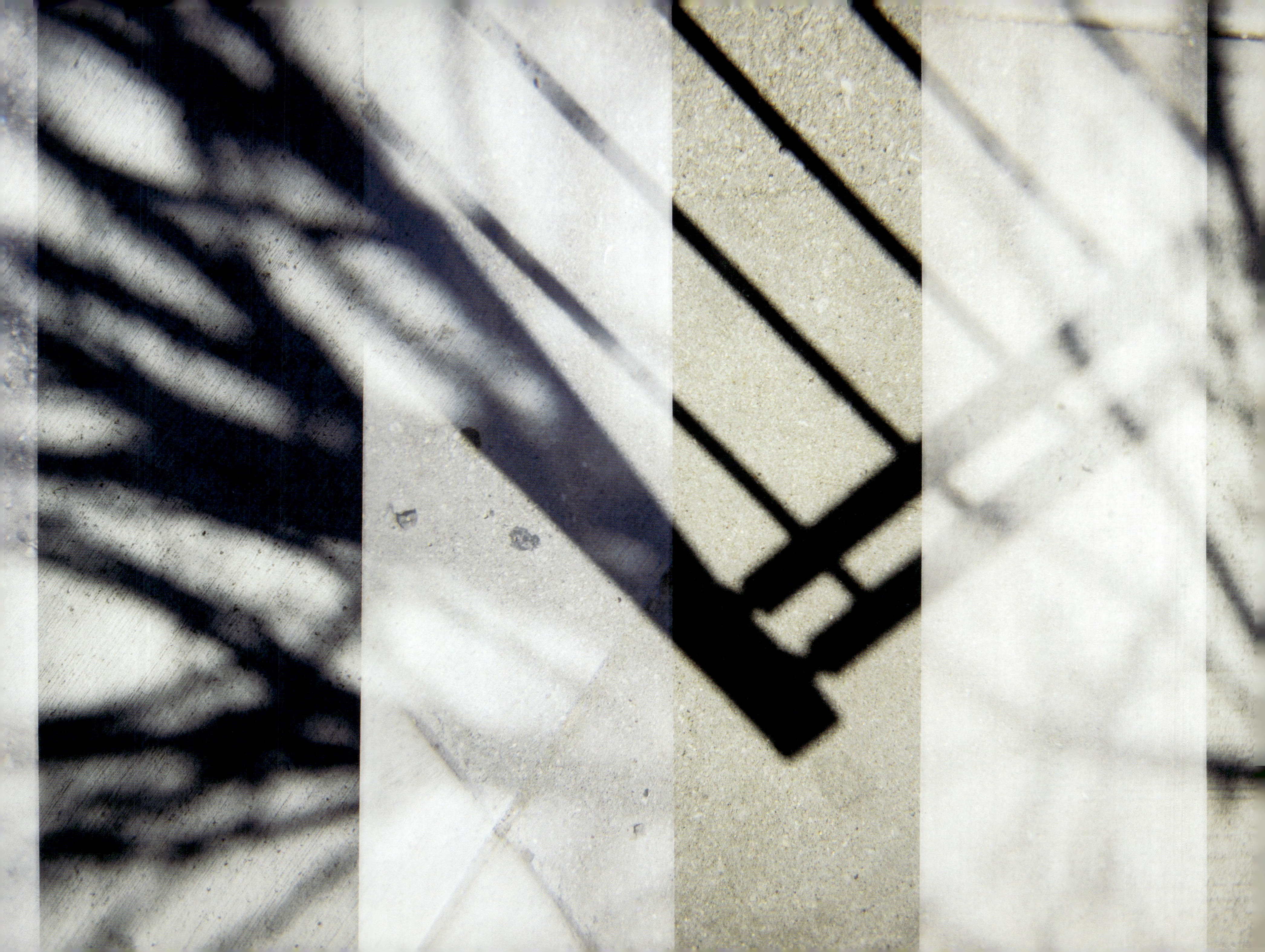

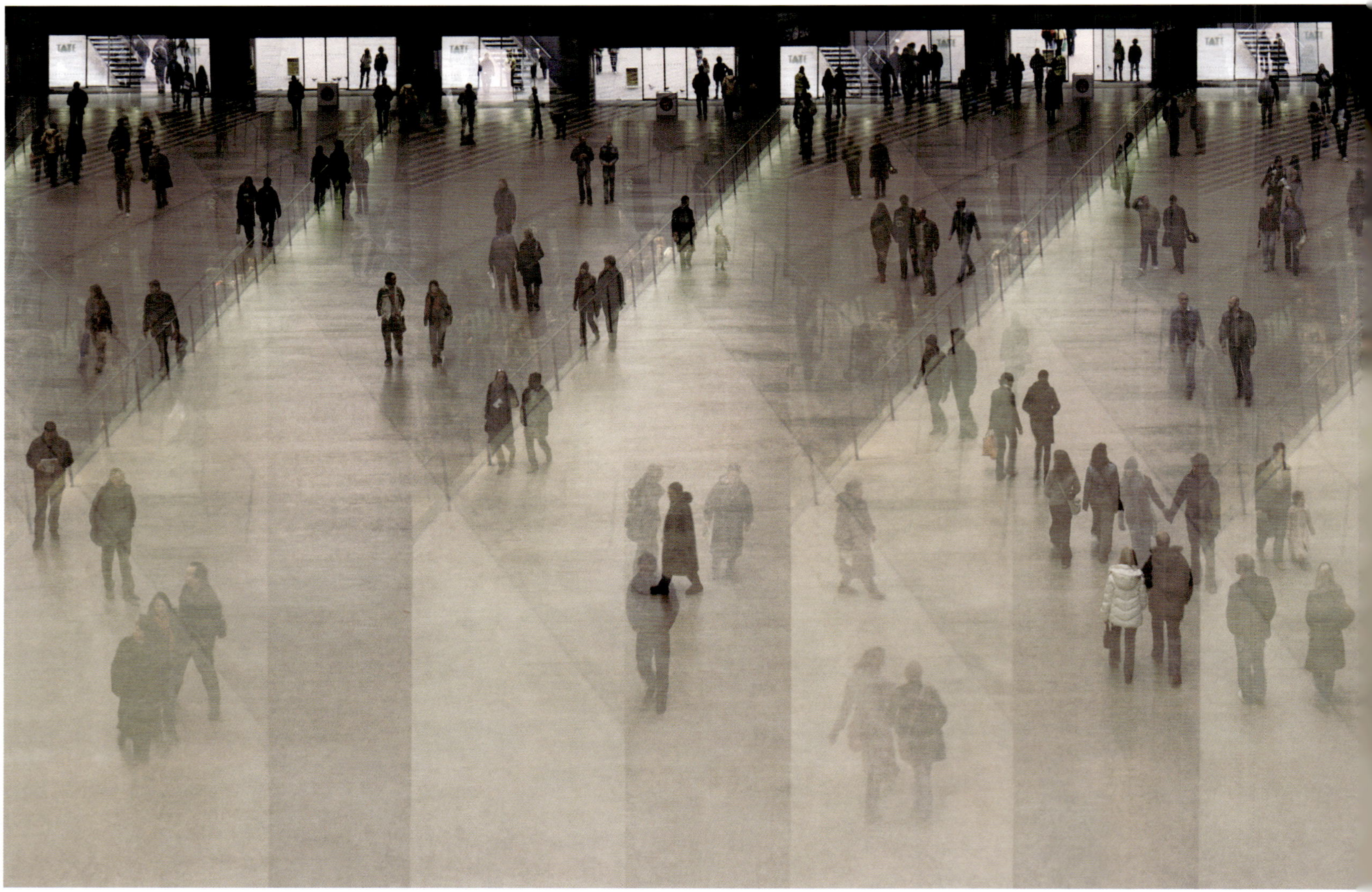

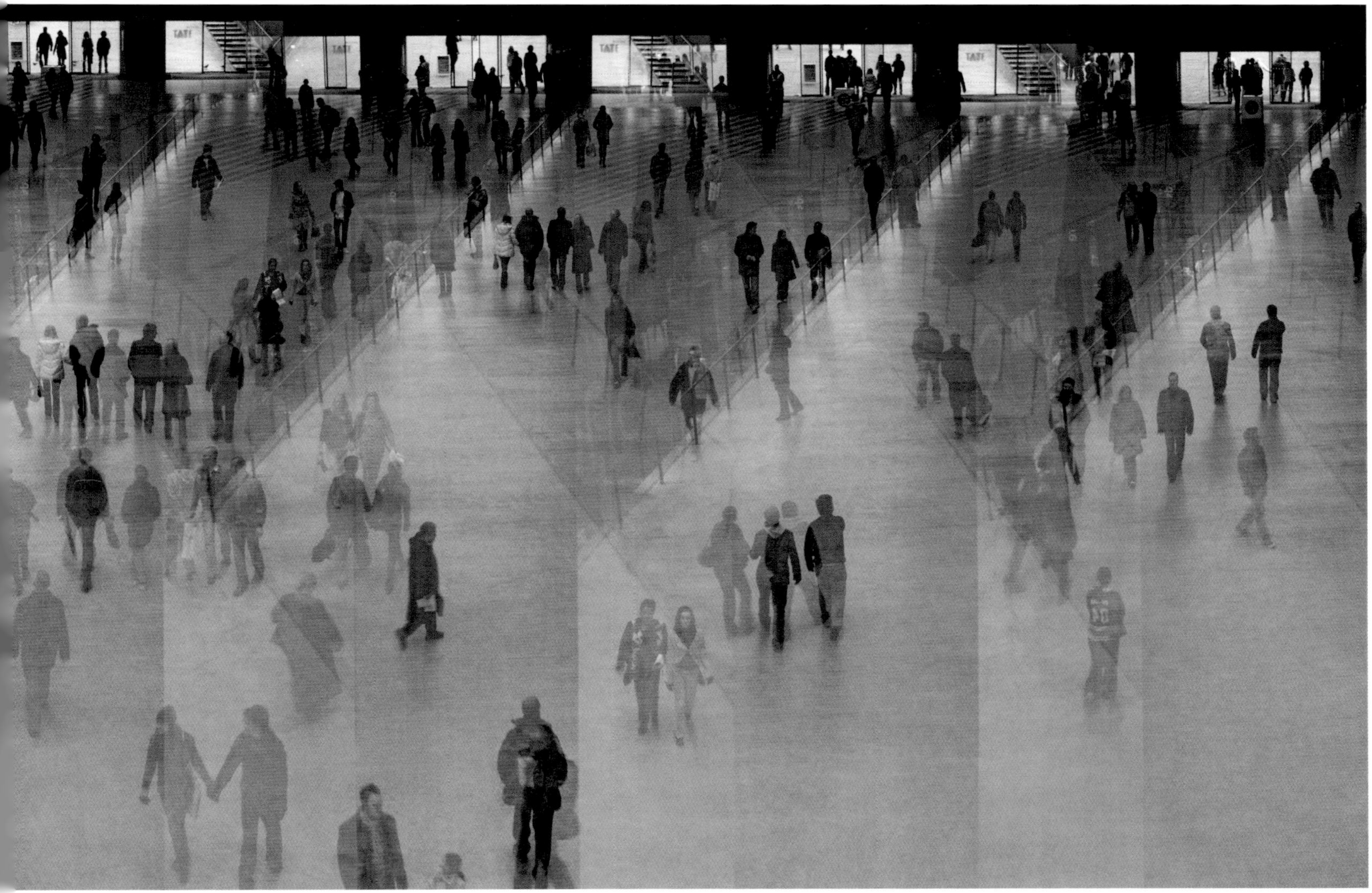

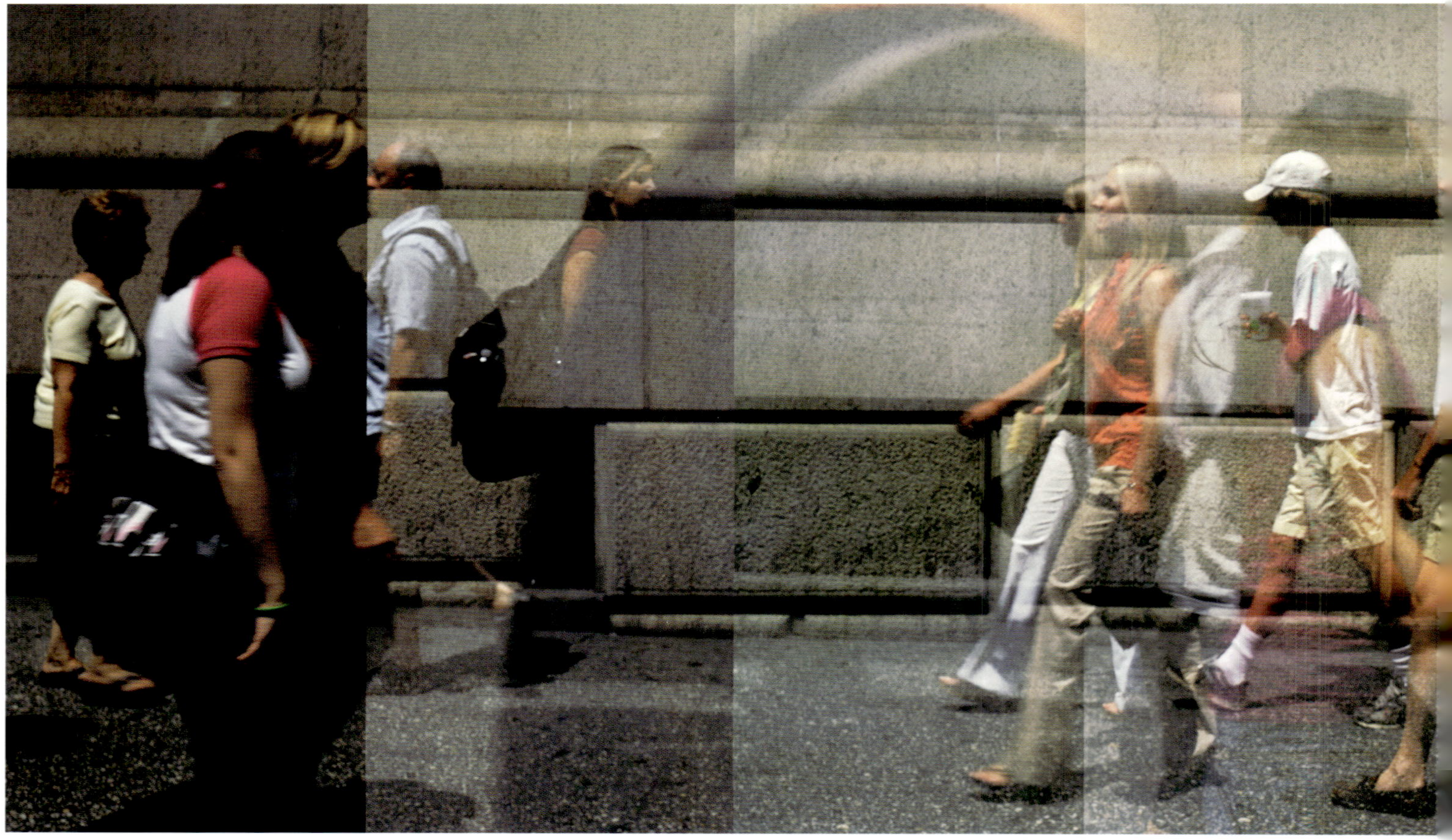

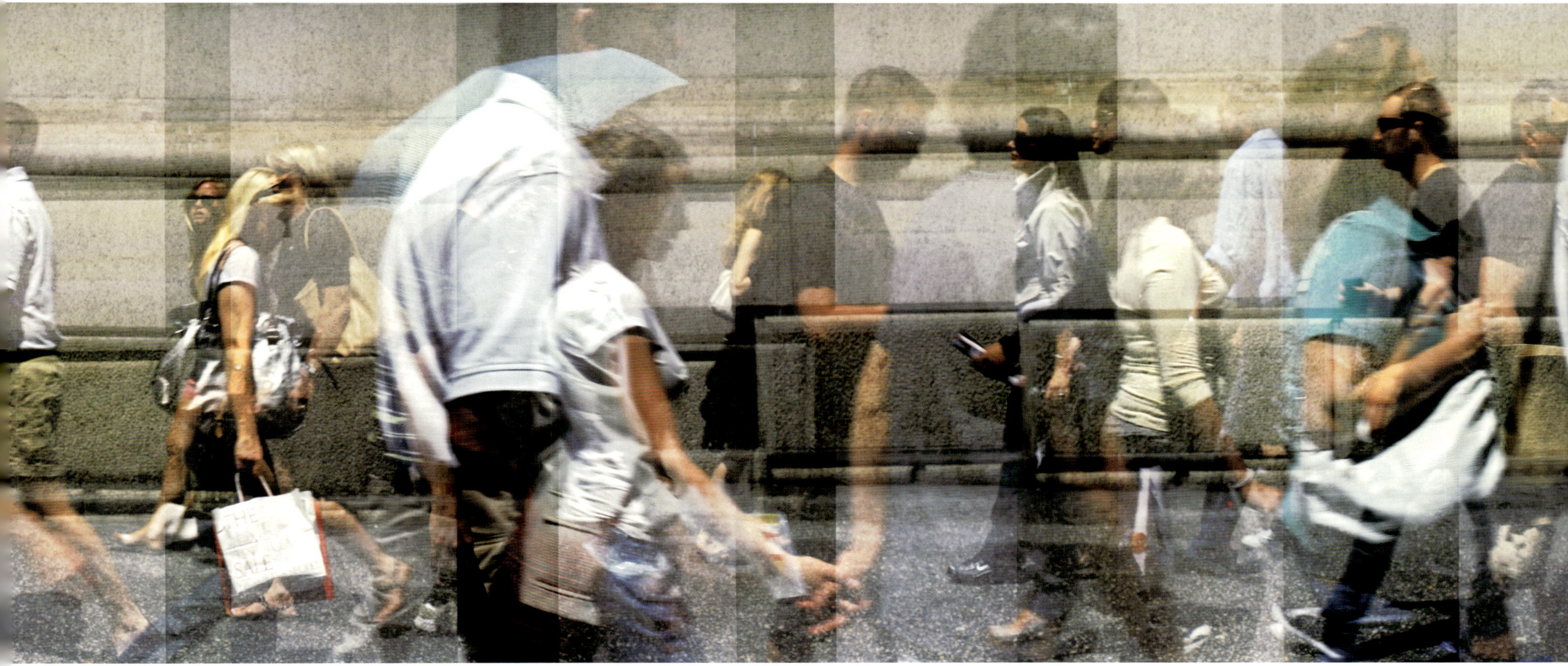

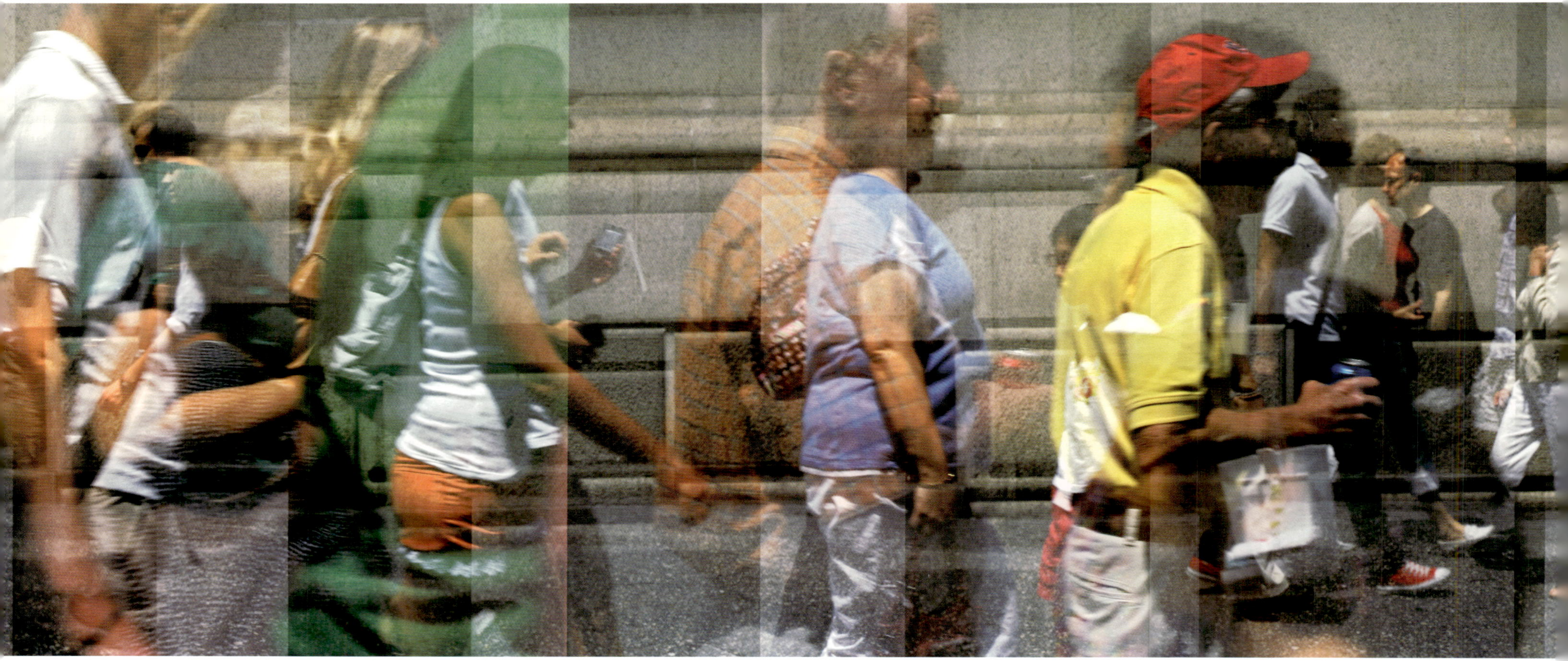

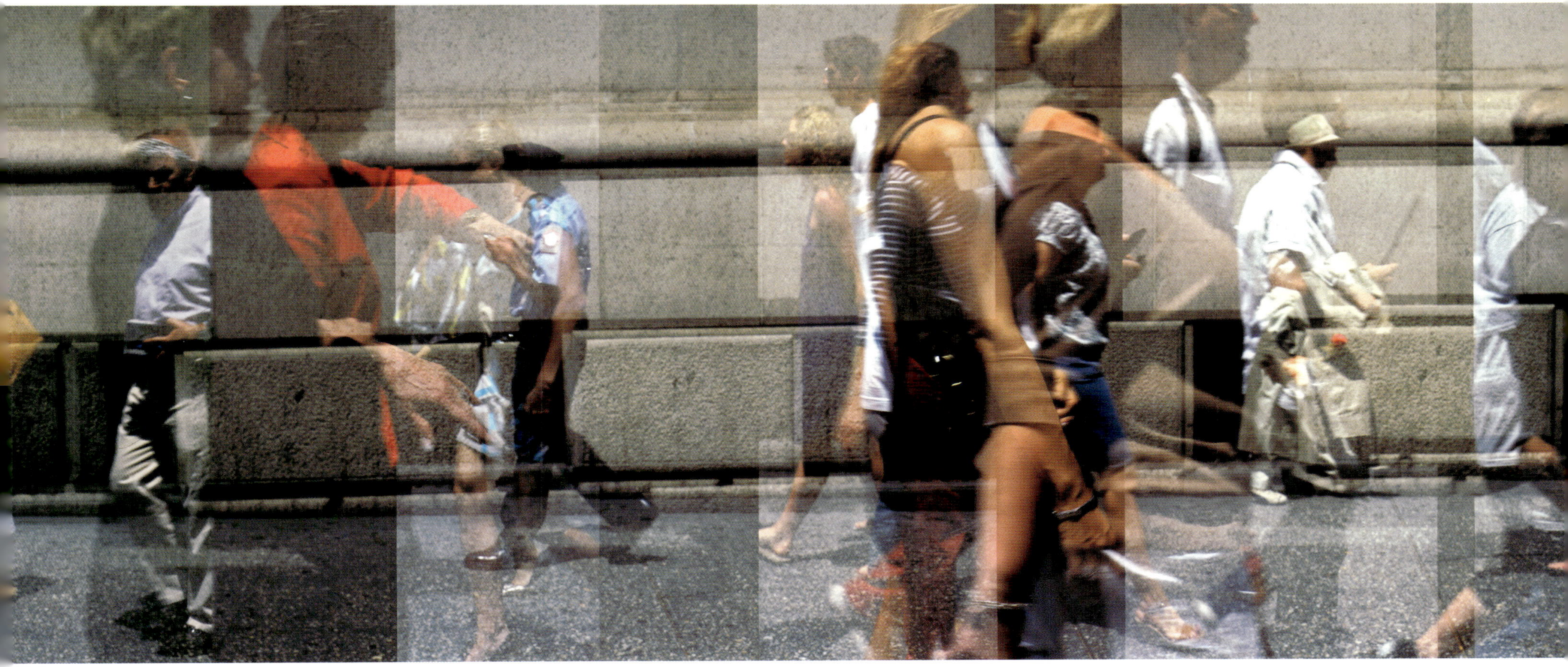

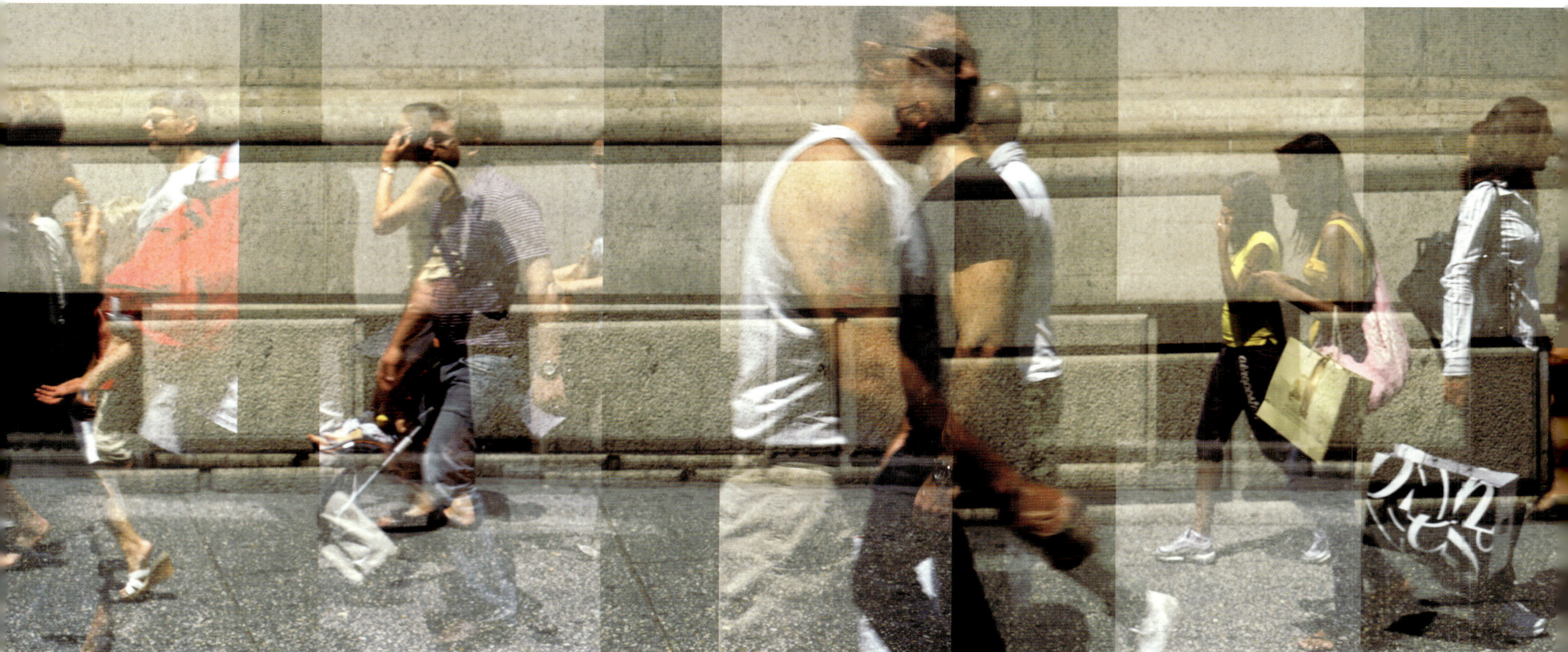

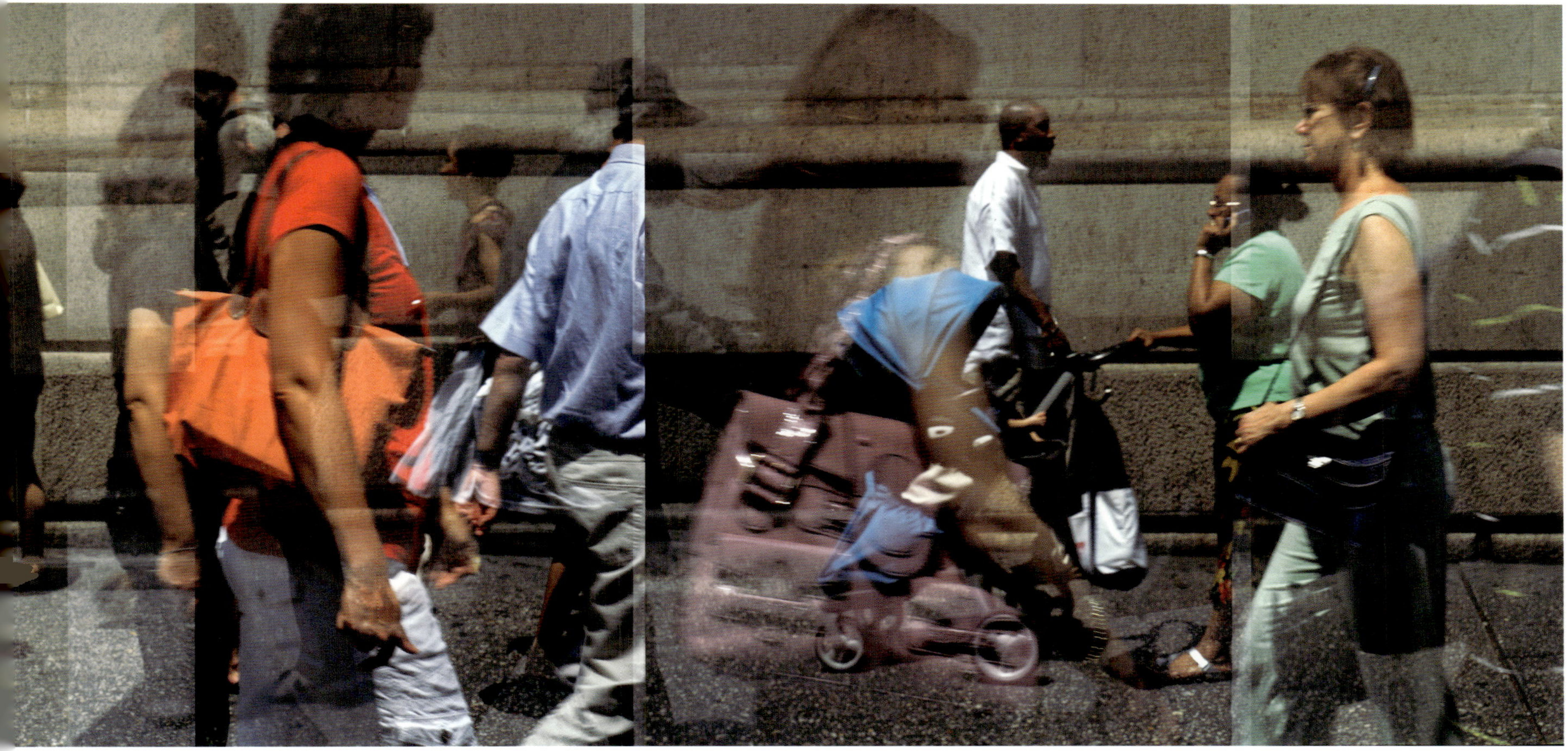

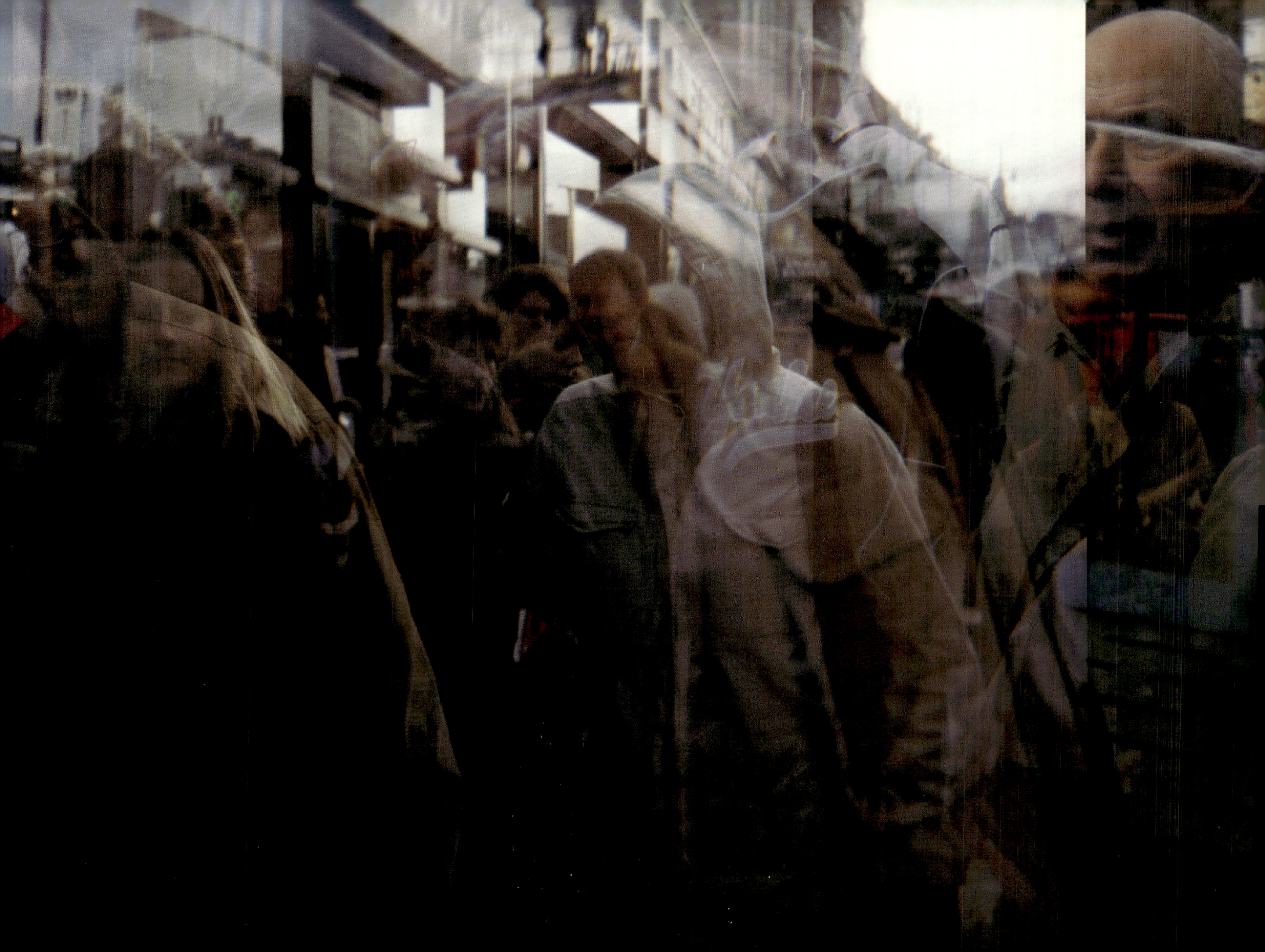

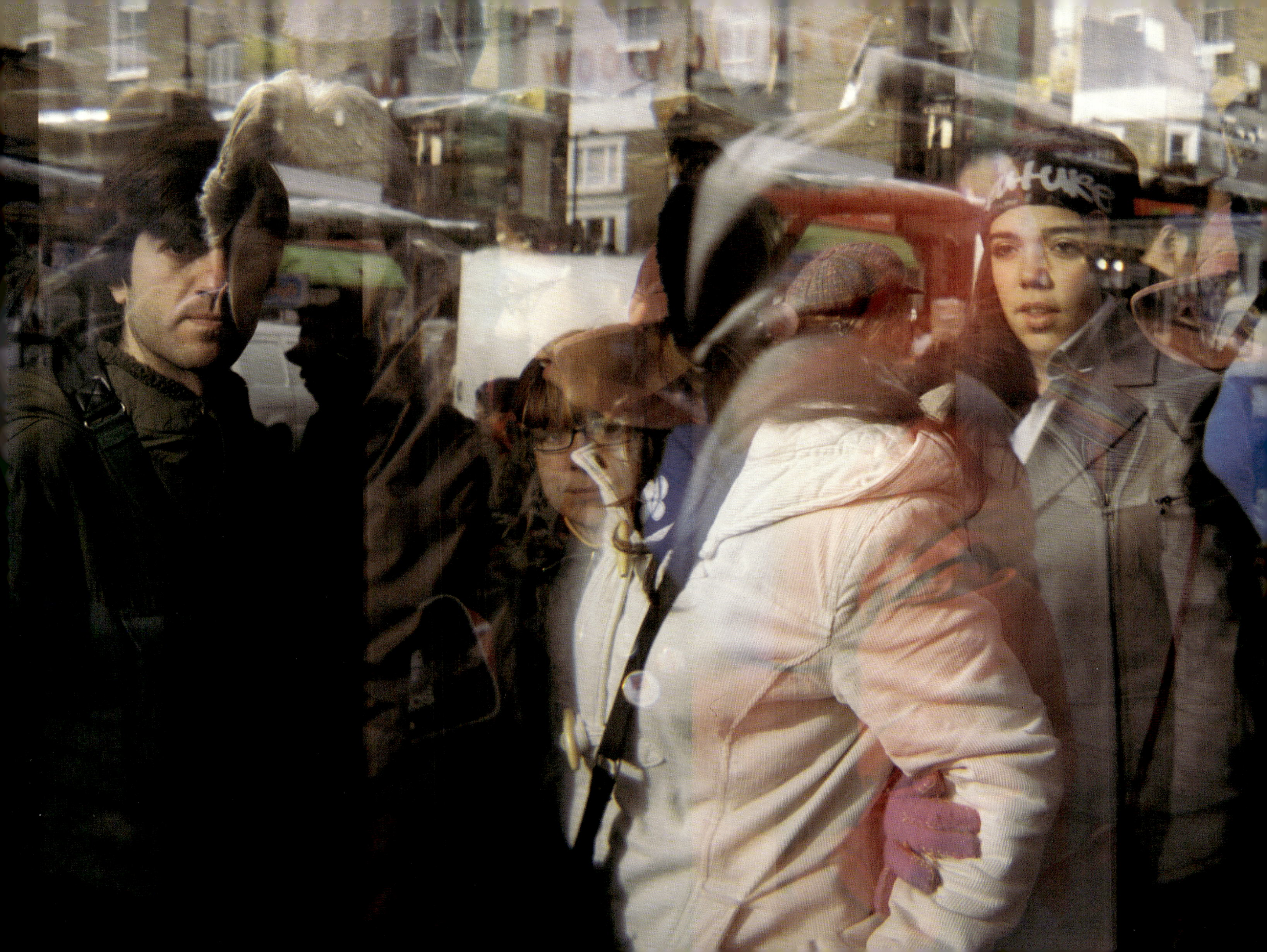

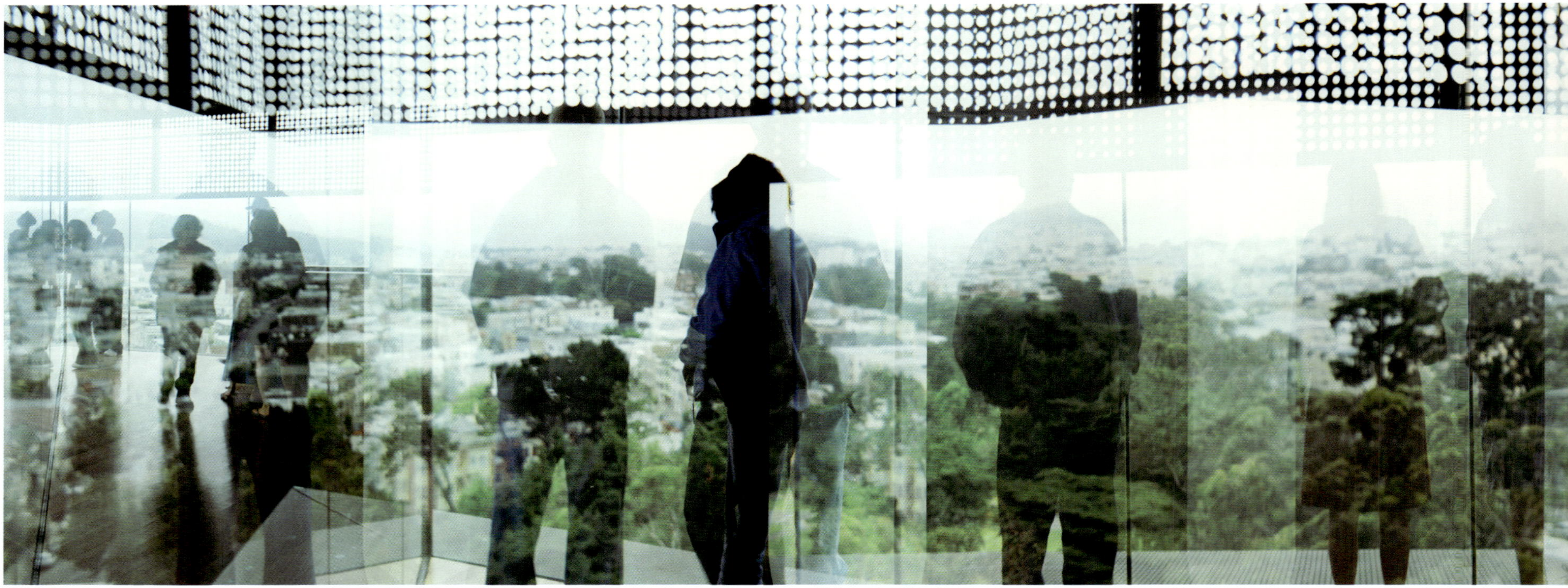

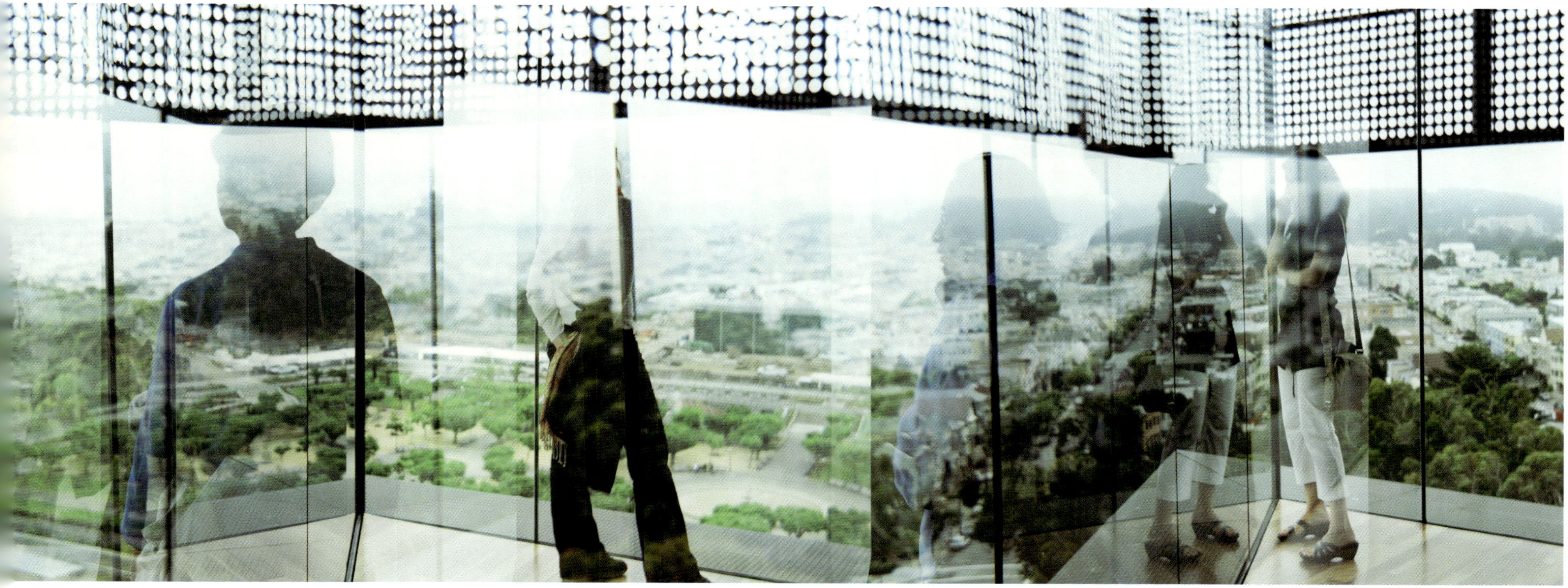

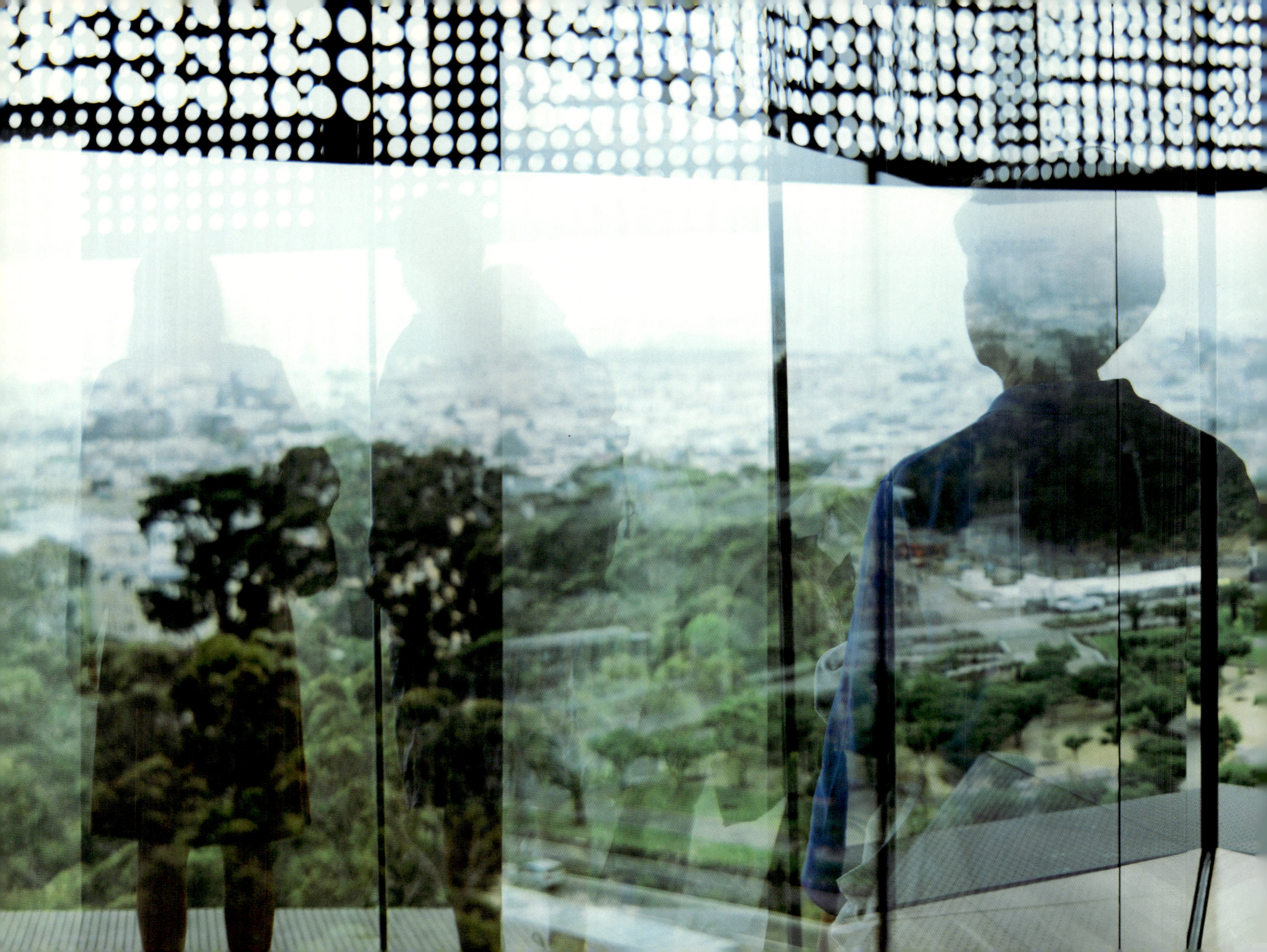

LUSH LIFE: THE OFTEN DISCONNECTED WORLDS OF HUMANITY AND NATURE.

by Bridgette R McCullough Alexander

In the universe of nature and man, what is the relationship between these two seemingly adverse entities? Can we say that humanity and nature exhibit any real oneness?

In the natural world, the dominant time signature is slow and nearly constant, and change takes the form of large cycles, or ebbs and flows. Birth and death flow into each other and what lies between is the exuberance of life. The nature that humans live and feel in, as opposed to the nature we know or understand scientifically, is our environment.

In our environment, the elements of birth and death indeed are present and recur, but we humans usually do not accept the continual process of birth followed by death. We deploy hundreds of tools to deny and avoid this aspect of nature, to control it or replace its truths with our myths. As a result, humans have reshaped almost all of nature to the point that throughout most of our globe we can hardly tell the indigenous species and original landscapes from those that were introduced by our ancestors or ourselves. So successful have we been at this in the past few centuries that life has become stagnant... apocalyptic, even.

Although we can no longer intuit nature with the clarity of the ancient worshippers of natural phenomena, nor even reproduce the mediated visions wrought by early Buddhists and the transcendental poets, we can work at it. On close examination of nature, we begin to understand its cadence. We can witness this in the elements which by all accounts seem to move within the laws of nature much more smoothly and, dare I say, respectfully than humans.

Douglas Fogelson's photographic studies accomplish some work toward this understanding. For example, we watch as trees pass through the process of the seasons—fall, winter, spring, and summer. At the opening of autumn, the trees begin to prepare for what will later come in winter. Their leaves dramatically and beautifully change colors right before our eyes, or as seen in *Exchange*, their internal energies move toward the long period of restoration and relaxation. This photograph presents the flow of nature, from the slow-vibrancy of yellowing green leaves to the interplay of semi-bare branches.

Let us look at *Exchange* to follow the truths of nature written smaller in the life-processes of trees. Obviously trees do not encapsulate all of nature, but represent only one perspective on nature—a metaphor—but that is how the photographic art itself functions, at least in part, metaphorically.

At the entrance of winter with all of their branches bare, the trees have moved within. During this time the trees take on a new self. The trees are barren, appearing to be static, yet the darkened branches and trunks resonate life at its most crucial moment in an inner place. What appears to humans as the season of death—and has been used for centuries by visual artists and poets to symbolize human death—the cold barren winter, the drought of vibrancy—is actually life internal, the internal process of a new life forming from death. These trees at this moment are in a revitalization stage and hold a magnificence that most humans can only see during other seasons.

It is during spring that the trees exhibit to us the process of birth, beginning with the budding of tiny green leaves. The branches of the trees and their trunks move from the protective darkened state to an outburst of green covering their visible body. Within weeks the trees evolve from semi-green and darker gray colors to another, more intuitively transparent vibrancy of greens, with branches full from the burst of leaves singing and moving with the wind.

The final stage for trees is summer, when their leaves achieve the largest size we will ever see them attain and they surround us in all of their outward magnificence. This seems to be the moment in the lifecycle of trees that humans ironically most connect with and most disconnect from. We spend a great deal of unconscious time with trees, really only engaging with them for our own leisure, for example, the summer picnic, the refuge of shade from the solar heat, the use of their branches for outdoor activity and their trunks as an easy chair on which to sit and read. But how little we understand the essential process that trees are undertaking at that moment. Trees are preparing for their next diminution and repose according to the laws of nature.

The Bounty of Nature

Each season I am privileged to witness these changes in the trees outside the windows of my home. Surrounded by lush and beautiful maple trees, I have watched their transitions and allowed their processes to become a part of my own birth, death, and renewal in kind. This is what Fogelson refers to through his work, as in *Bounty*. Bountiful is the full-on rich, lush, taking over of each frame that he lays together. In that photograph, it is impossible at first glance to see where the frames begin or end, much the same with the beginning and ending of the season-cycles of trees. They are all so intertwined with each other like the relationship between human and nature. Just as human and natural elements recycle into each other, as we re-enter the earth and the earth returns to us, our essences merge.

Just recently I have begun to incorporate the arboreal laws into my own life, finding comfort in the processes of trees and knowing that existing within myself are these same elements and processes. Here are some of my thoughts as I find them reflected in Fogelson's studies.

Life seems to be always between the natural ebb and flow for humans, the flow of life and then the abruptness of death. How can we find the harmony between this diametric opposition that nature, seemingly, has accepted and even flowers from? Some intuitive answers to these questions may lie within the photographic study that Fogelson calls *Biota*, which I will later discuss. Fogelson's works, in particular the beautiful *Bounty*, take us into the worlds of both nature and the human environment, and allow us to see them as elements that are not too very different.

In Fogelson's studies, the process of time (or reproduction) is quite the central theme, one that I interpret as *hybridity*. This concept appears frequently in today's eco-friendly movements. Here, I am using it to mean the connection or reproduction between humans and nature. Nature has been "mastered" and shaped by humans—we have learned this to our shock and horror—but how can the human be reshaped by nature? Can we choose a more "natural" way of life?

One way this can be done is to re-learn how to see the beauty of nature and choose to live in a harmonious way with nature and other humans. Many teachers have defined this path. One critical feature here is that we recognize the beautiful intuitively—as the beauty of nature and the human are the same (or very close).

Another less familiar path is to face the differences or dissonances between the human and nature. The German philosopher Immanuel Kant referred to this difference in his concept of the sublime beauty of nature.

Fogelson's studies might help us interpret the sublime as more of a hybrid force and less of an oppositional reality. *Bounty* brings us very close to the natural ebb and flow of tree life, a process where "continuality" is present. The method of photography Fogelson employs is similar to the continual process that we find in nature. Fogelson overlaps photographic frames, and in this process creates for the viewer a new continuous and connective element within photograph.

You can see this very clearly in *Canopy*. Trees reaching high into the sky give a sense of protection in nature for humans. If *Bounty* gives us both positive and negative space, *Canopy* takes us to the next level by exploring inside those spaces with the film and/or camera. Fogelson gives us the tension or excitation that rests between birth and death...life. Life is what we see in *Canopy*, which celebrates variety and nuance at the top of an ancient jungle vegetation scene.

The leaves in *Canopy* take on an inviting three-dimensional quality, appearing to reach out and touch you. Meanwhile the shadows created by the overlapping of frames give us remnants of the history of these trees. Fogelson does something unique, instead of capturing or harnessing nature through his lens, he beautifully gives us the flow of nature by crafting his camera work, moving the images through the process of aiming and advancing his film. Nature here is powerful and mighty, all the more for its *persisting* beyond the photographic frame, a metaphor for perseverance beyond the spans of our puny lives and ambitions.

Biota

To be sure, these are photographs—a human creation and a human capturing of points of nature. But what brings a different reading to these photographs is the manner in which Fogelson actually makes the photographs. The processes by which we engage in the depiction are revealed in the spaces where the photographer's method of advancing the film overlaps and creates tension between film and subject. In frame after frame, Fogelson wrestles and brings forth two realities: both singular photographs of plants (or people, clouds, water) in different perspectives and moments in space that reflect the continuality that is common in the natural world, against which we humans struggle so vigorously in our environment.

Fogelson brings these two often-distant worlds together brilliantly without the appearance of overcompensation. In an example from the series entitled *Biota*, wherein *Canopy* and *Bounty* also appear, the work *Beverly Hills* stands out as a wonderful example of man and nature working together. It is indeed ironic when you consider the culture of this tony community to be the farthest point from anything "natural." Here, however, the imagery takes us to what may very well be the core of this culture, the discovery or rediscovery of beauty in the form of the all-natural.

Fogelson layers various stimuli over the frames and creates new images for us to reflect on nature and its presence. The plants in this photograph look almost plastic, a different version of hyperrealism in nature. In *Beverly Hills*, we are introduced to another aspect of nature, and that is humanity's relationship with it via control. At the margins of the photograph, we begin to get a glimpse of other "natural" elements, for example, the sloping of what appears to be a hill and also the sky off on the far left side of the photograph.

After decoding the developments in *Beverly Hills*, I am introduced to another exercise in both *The Tree* and *Sum. These* images serve as great reminders of where I began this essay—the connection or disconnection between nature and man. How it is that other elements and organisms in nature seem to understand and respect the natural flows of life and death (and of course everything in between), while humans seem to be at such odds with them? I suspect this tension existing within humans regarding nature is caused by what we find on the surface, what we can actually see, feel, and taste. And although I suspect other organisms and even elements of nature function very similarly under many of the same superficial rubrics as the human, there is one exception: nature embraces the ebb and flow, life and death, as an ongoing process, and we humans have not quite grasped that concept or at least that value.

One might try to hold onto the structure of Fogelson's *Flinch*, which is still beautiful nature, but a bit blurred at first sight. It takes a while for one's eyes to adjust to the layout of the work, but once that happens, a new understanding begins to take hold. As our eyes move out (naturally) from the center of the photograph, other elements begin to take shape—color, light, shadow, and focus. This work traverses the breadth of elemental nature, perhaps even forming an internal connection that humans have with nature, in a manner that brings a number of the pieces in this series together.

Through his lush and layered photographic studies, Fogelson brings us to a new, different, understanding, whereby we feel less existentially threatened about nature and more challenged to keep up with it.

organisms

in nature seem to understand ar

nd respect the natural flow of life

and death

(and of course

everything   in between), while

humans seem to be   at such odds

with it.

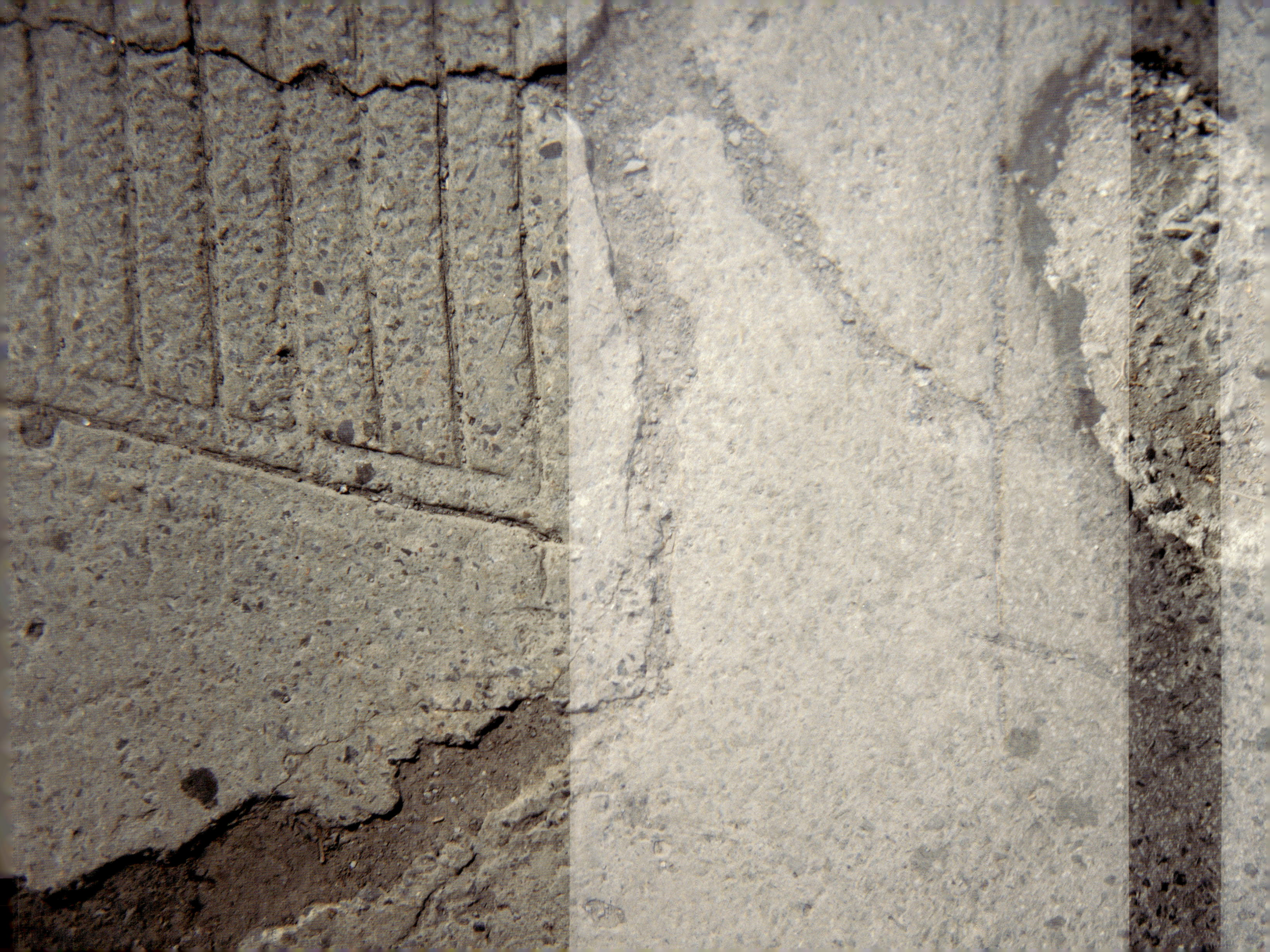

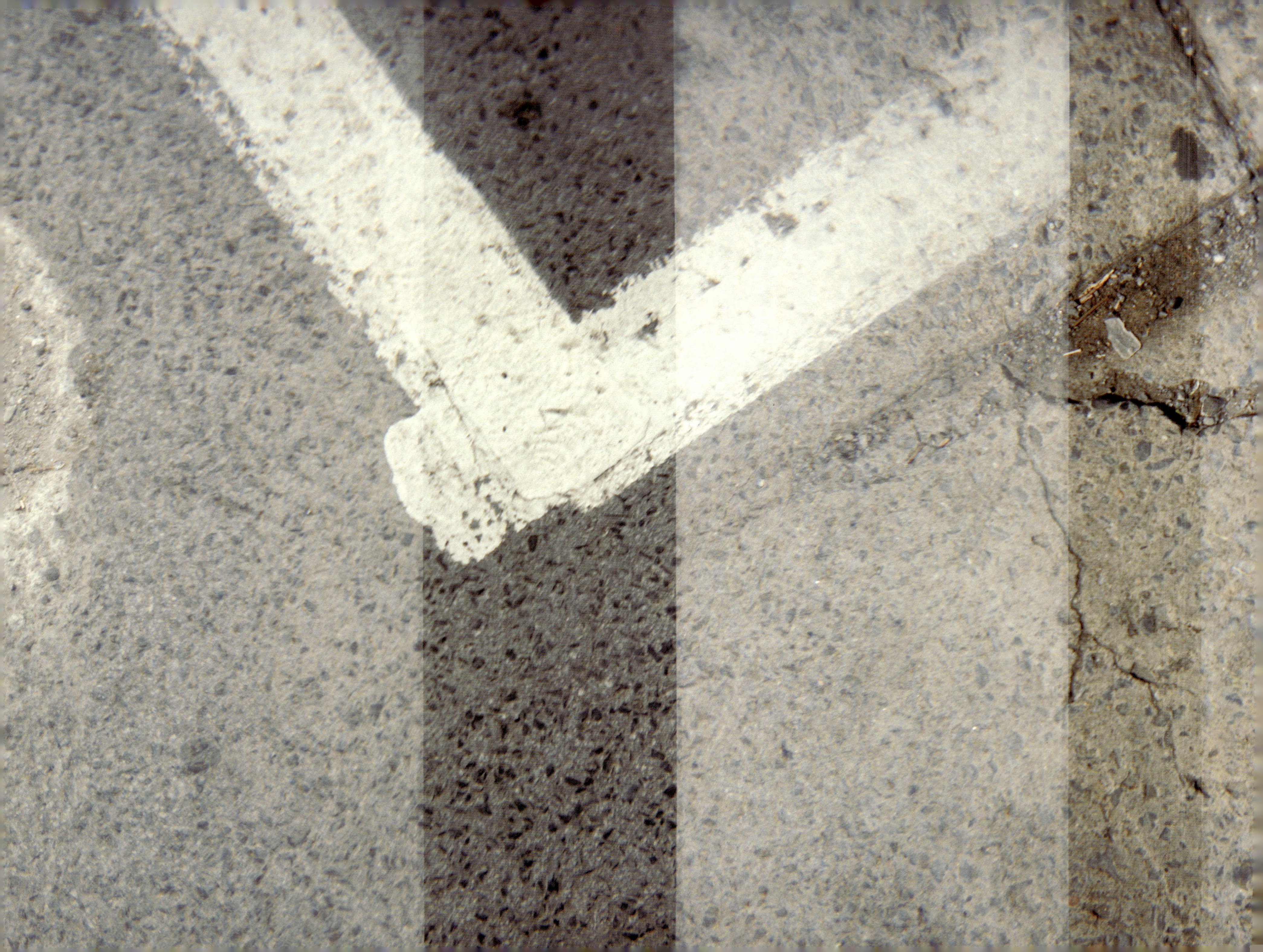

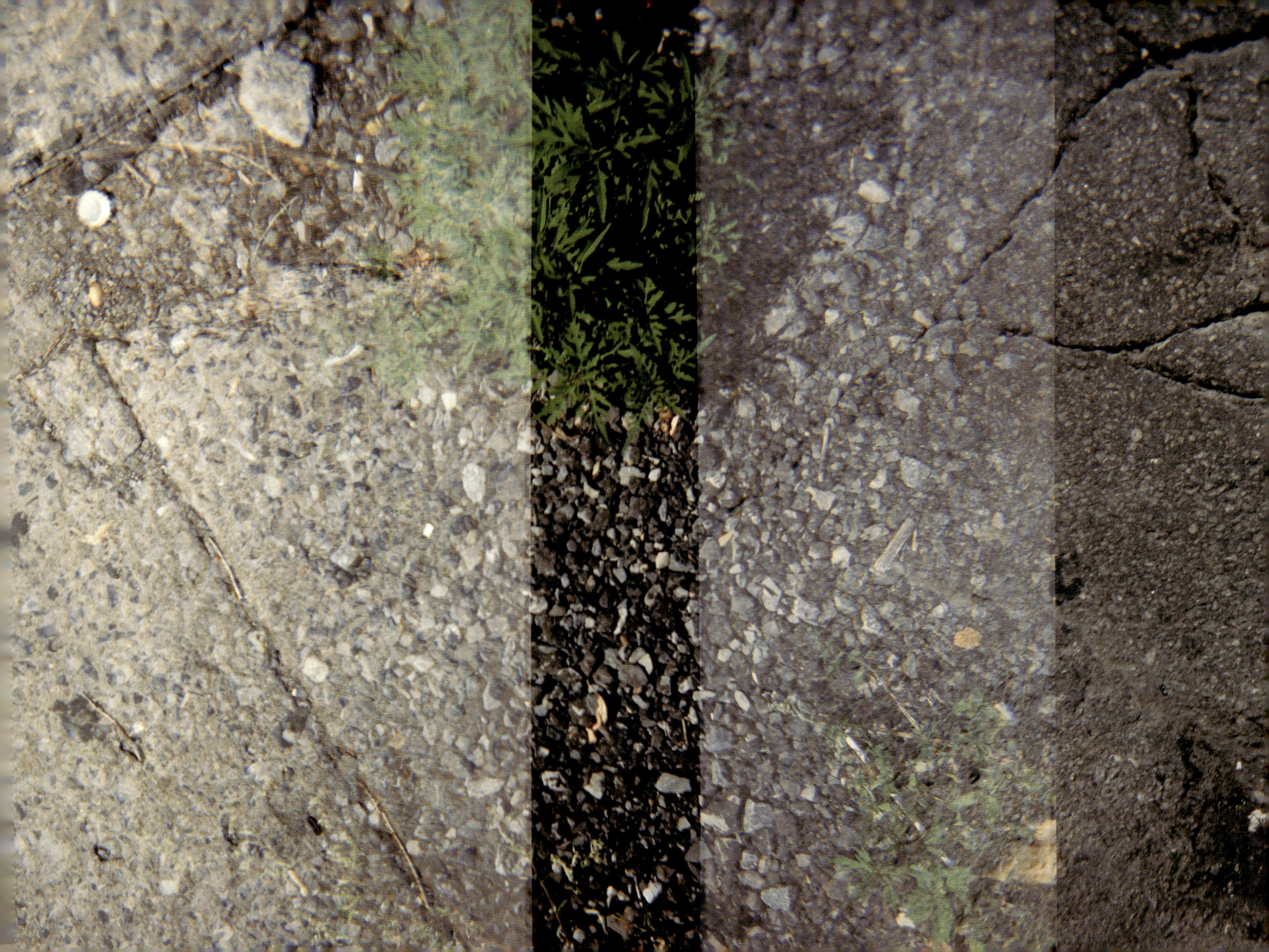

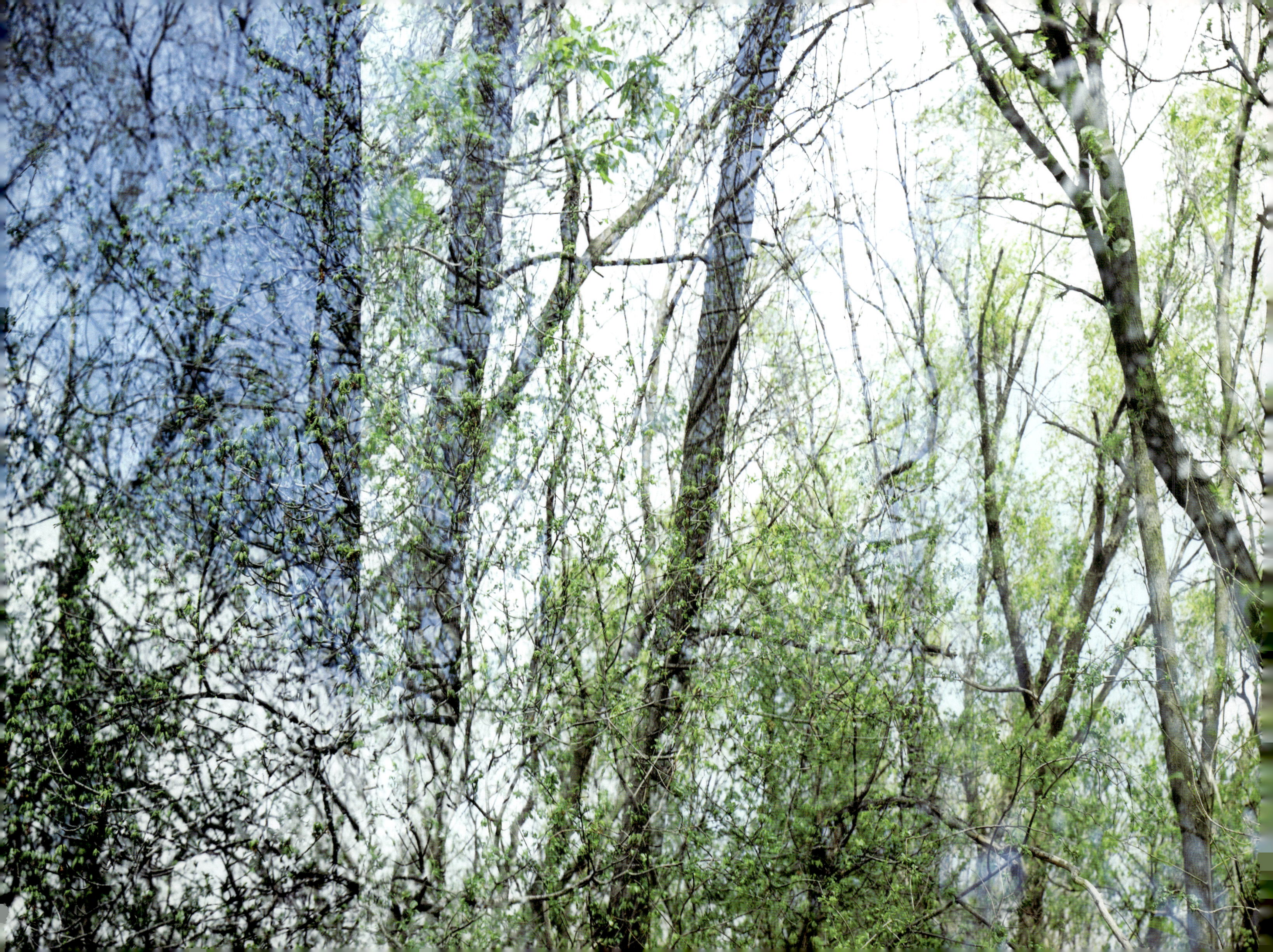

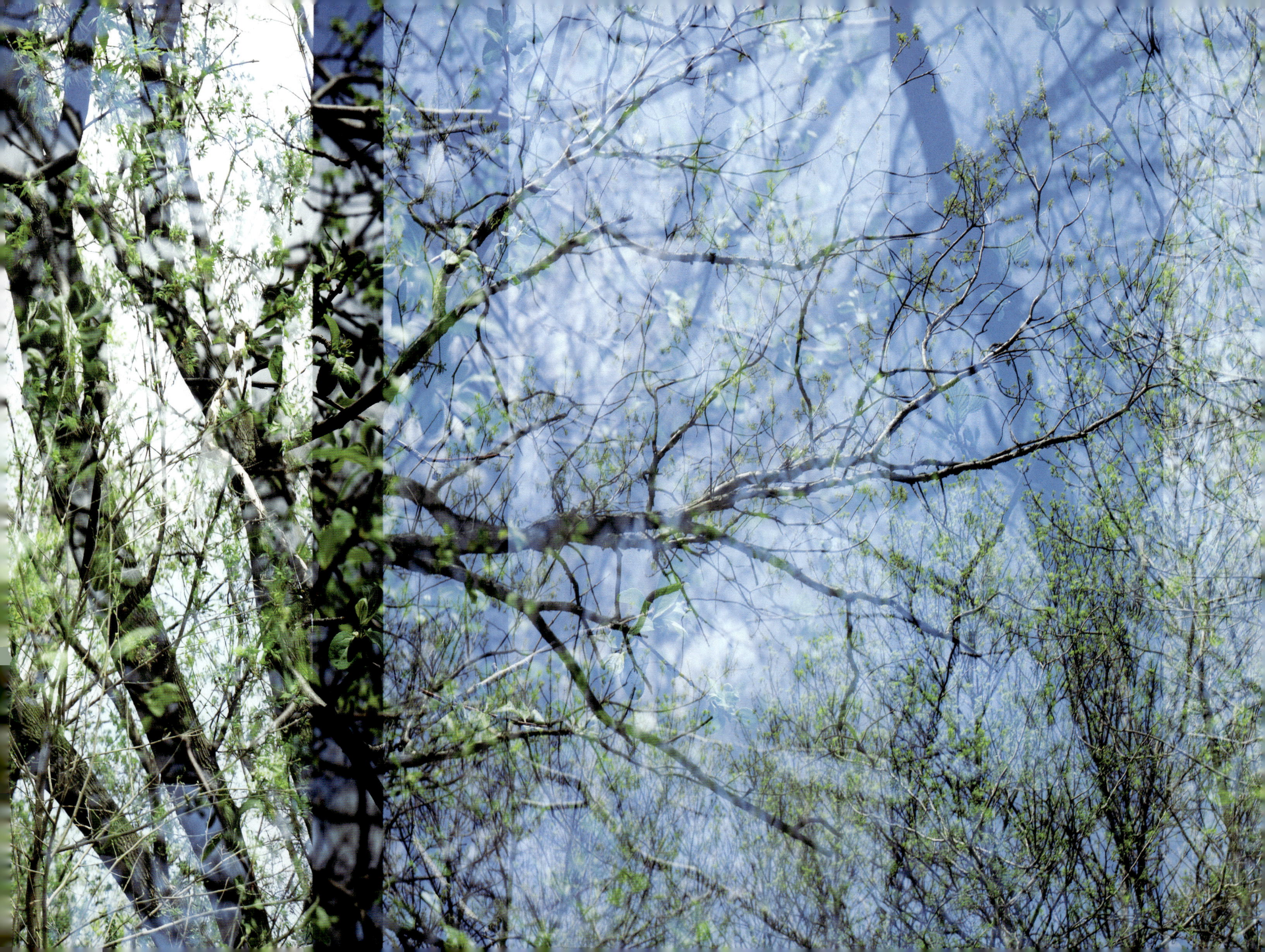

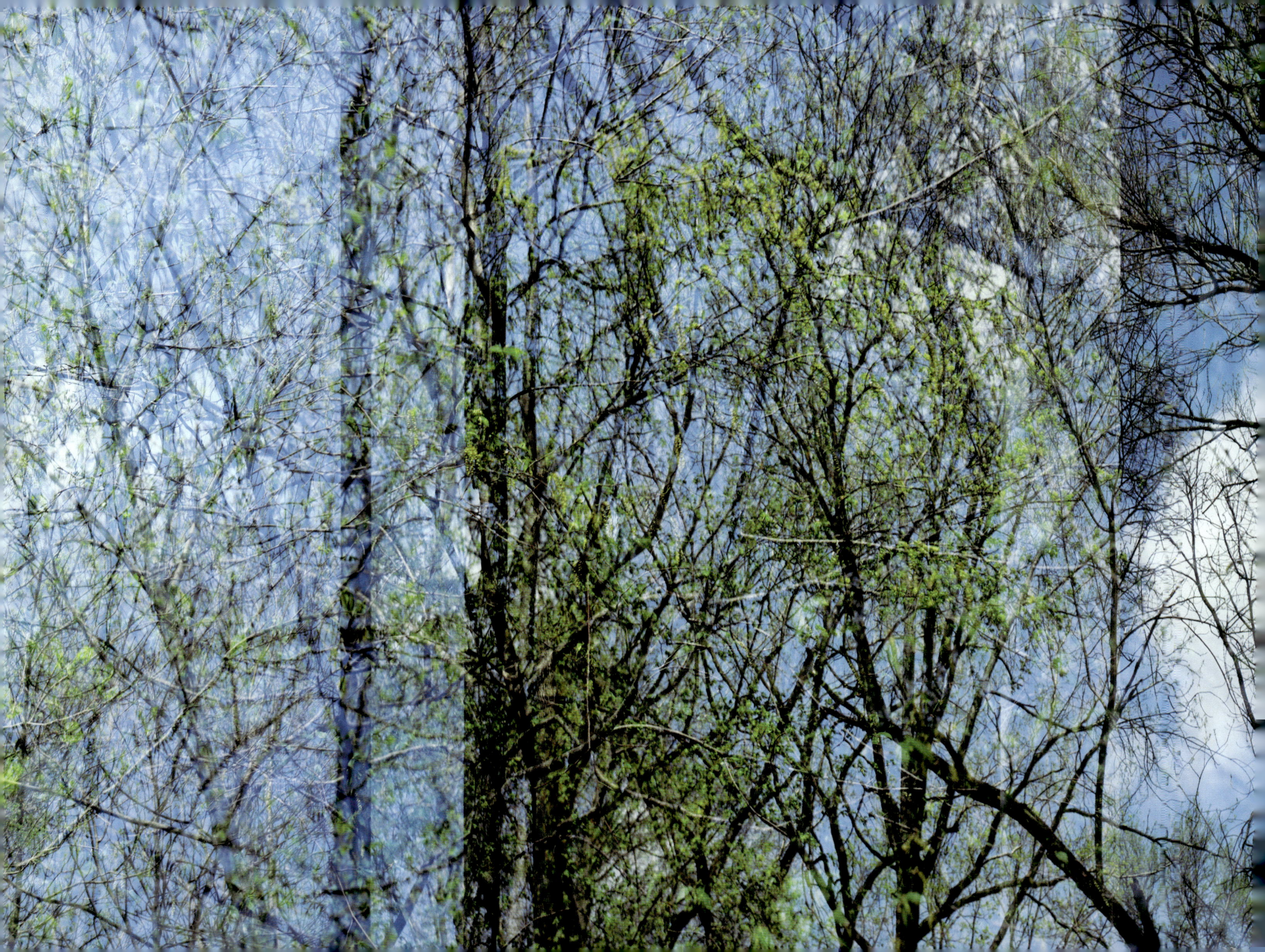

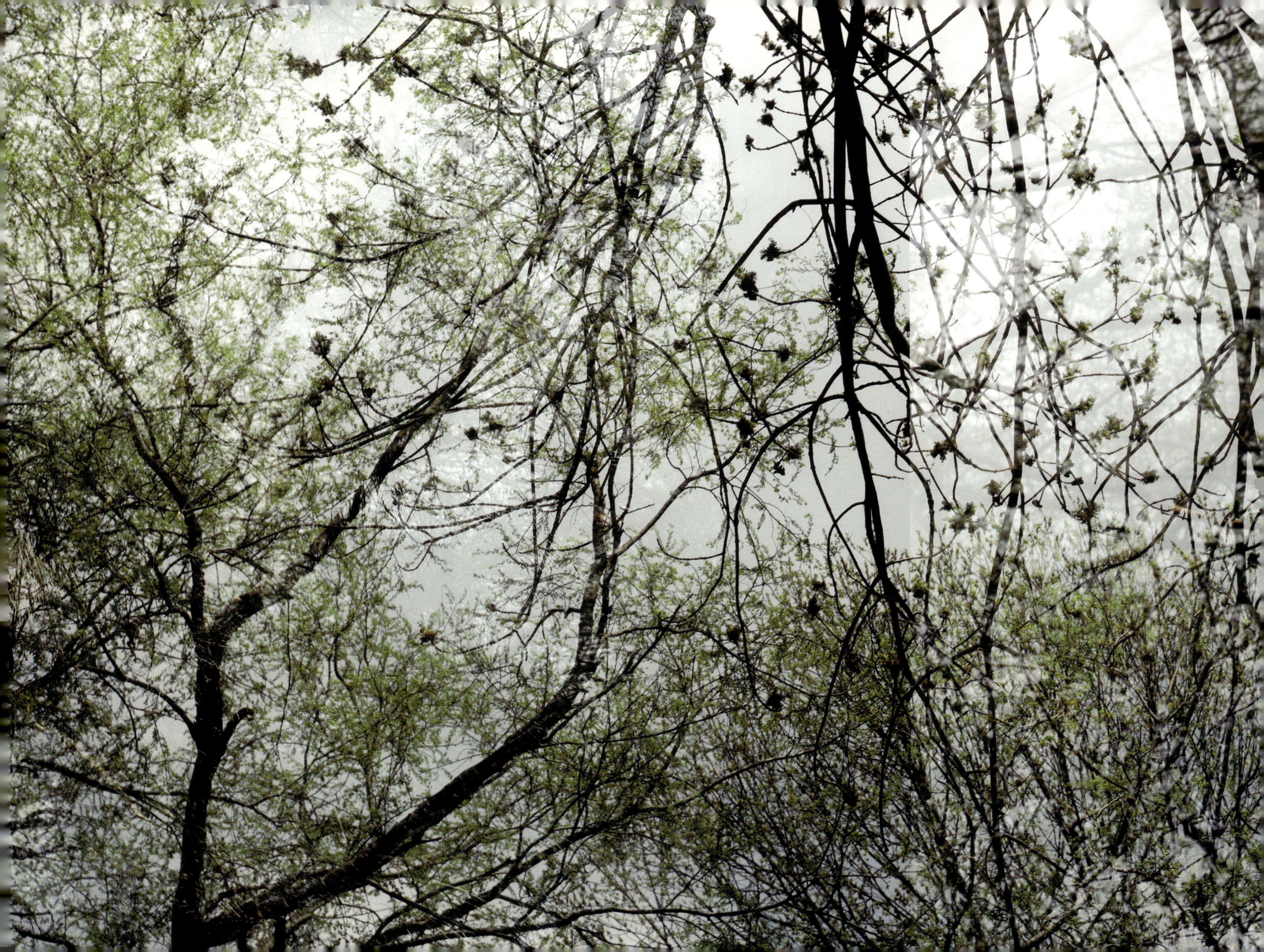

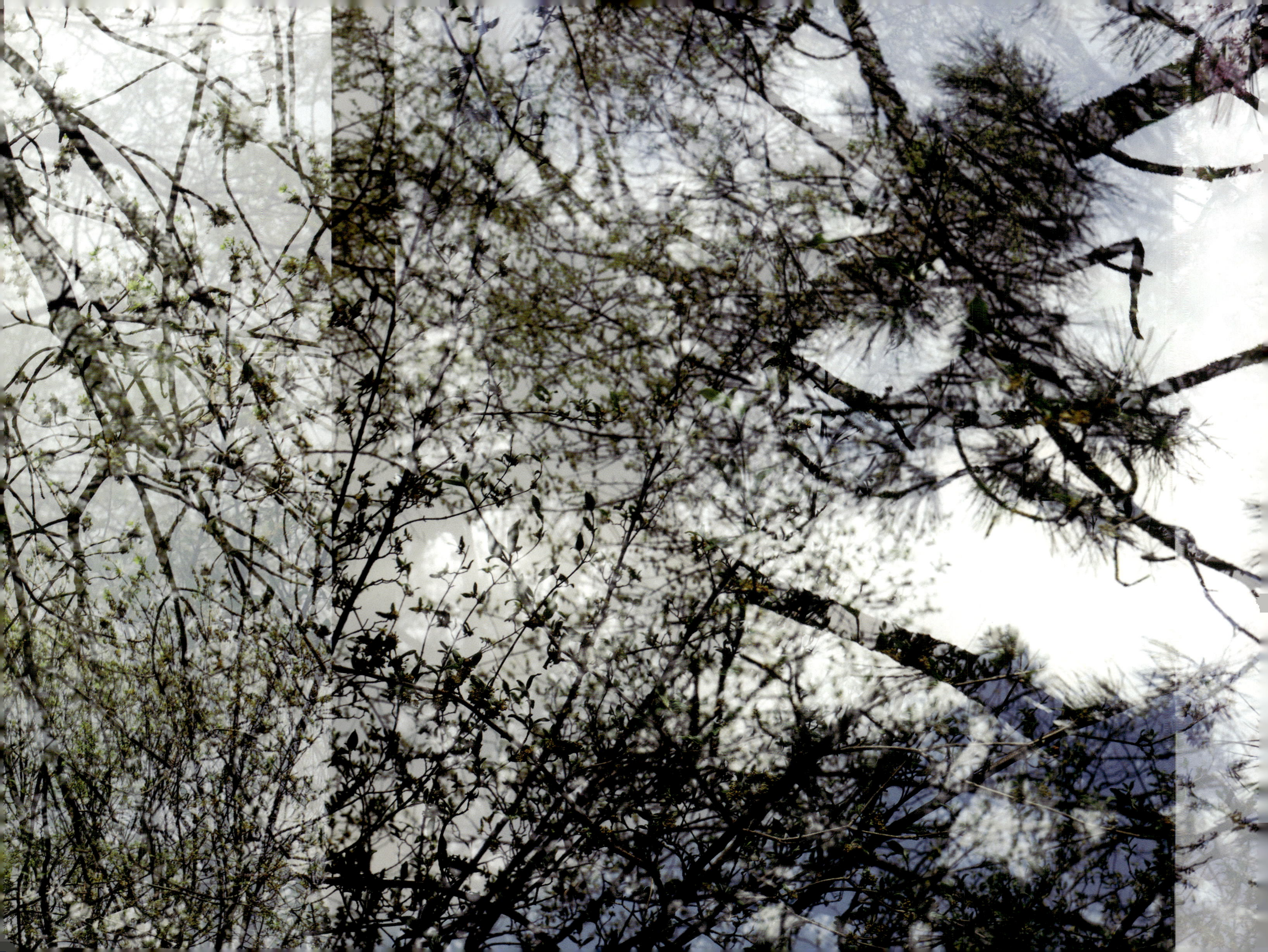

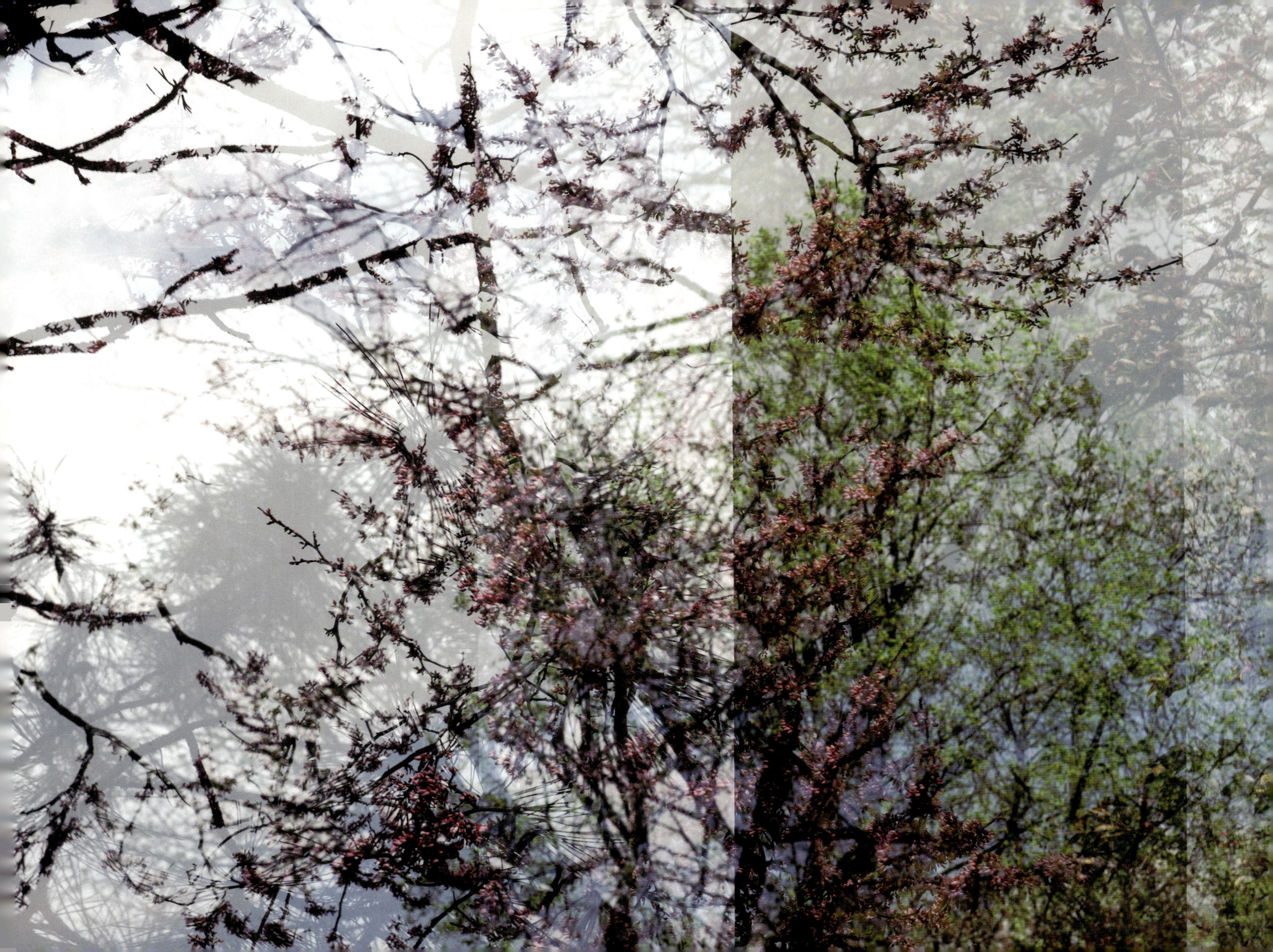

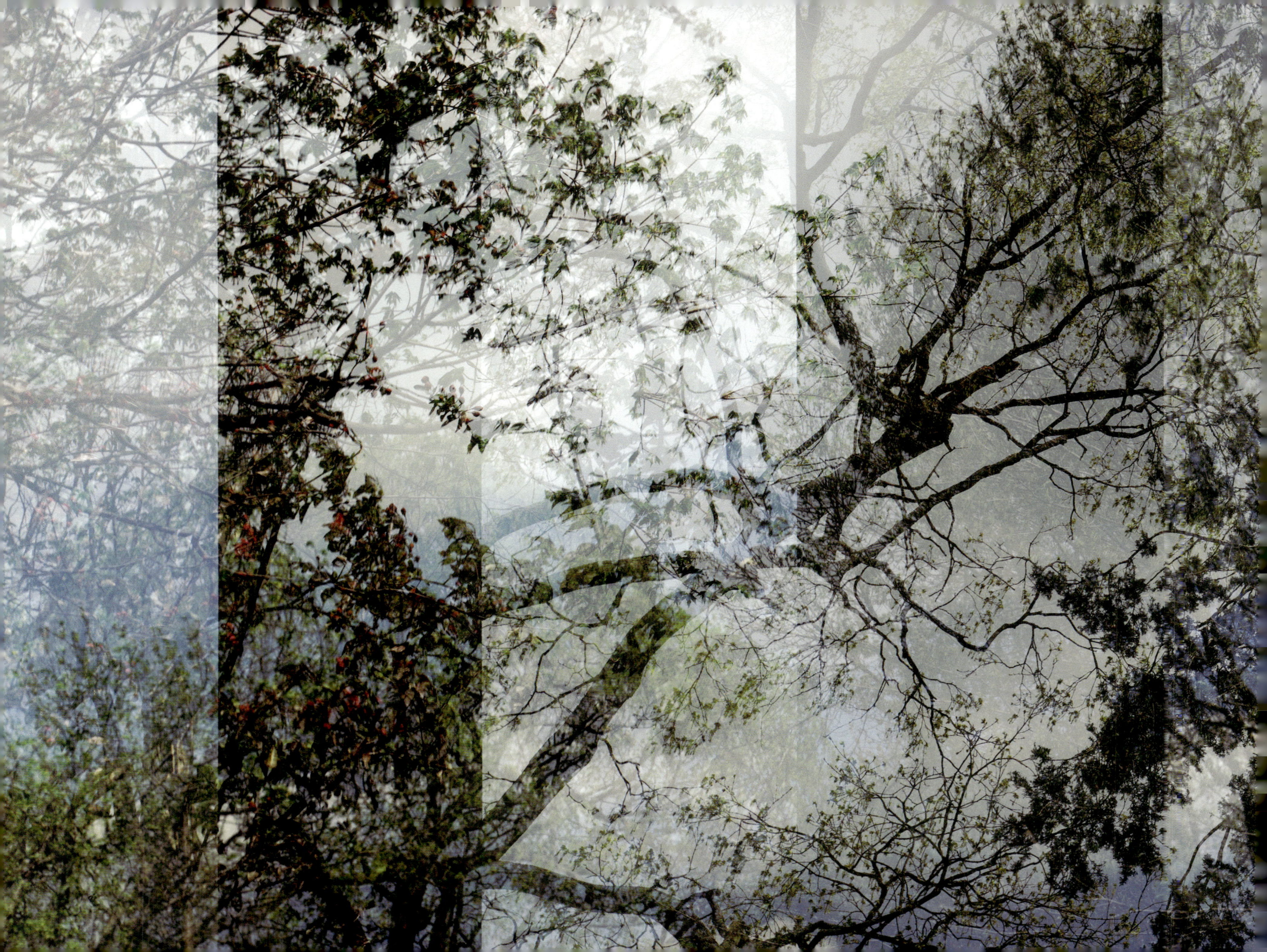

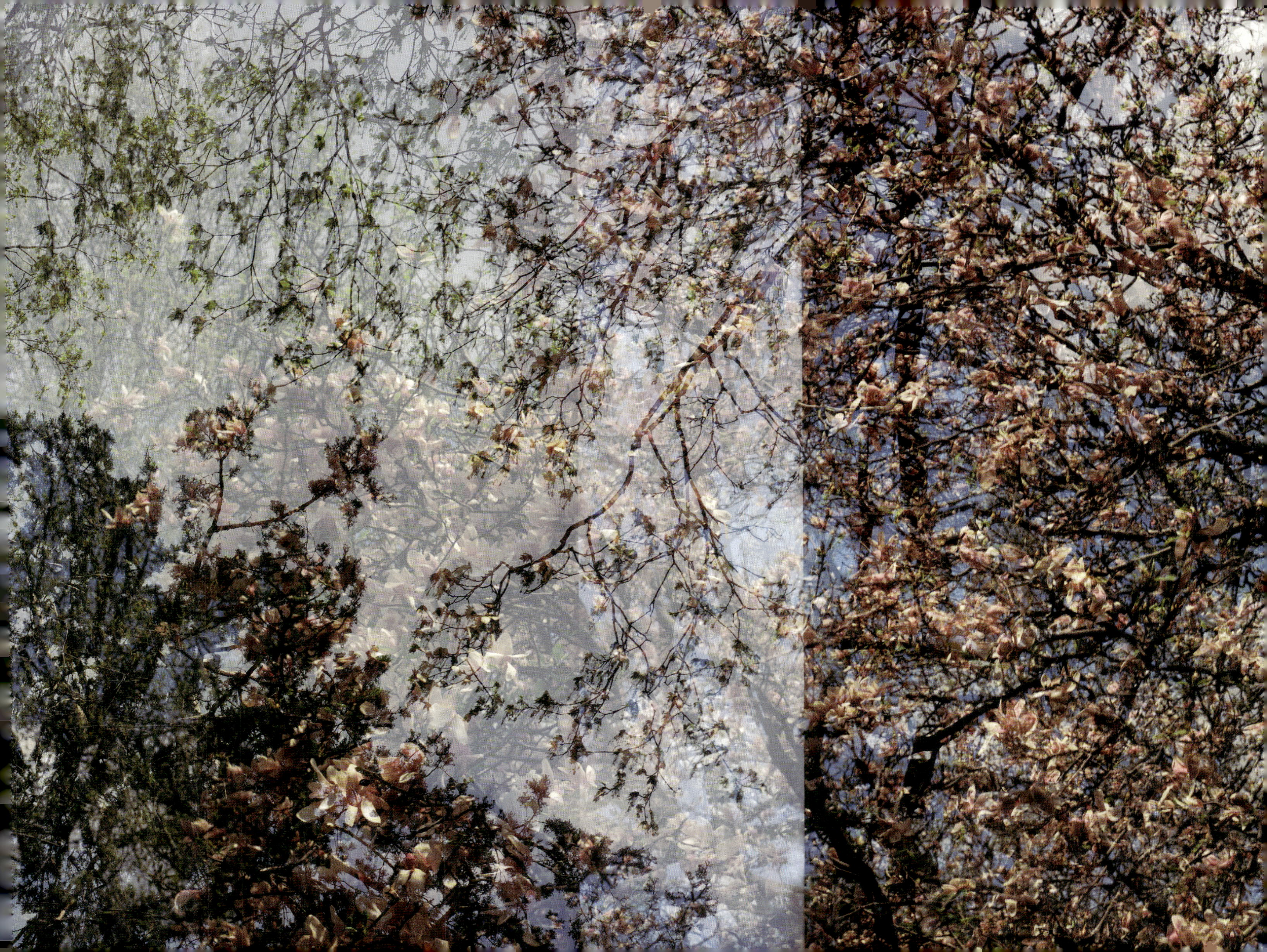

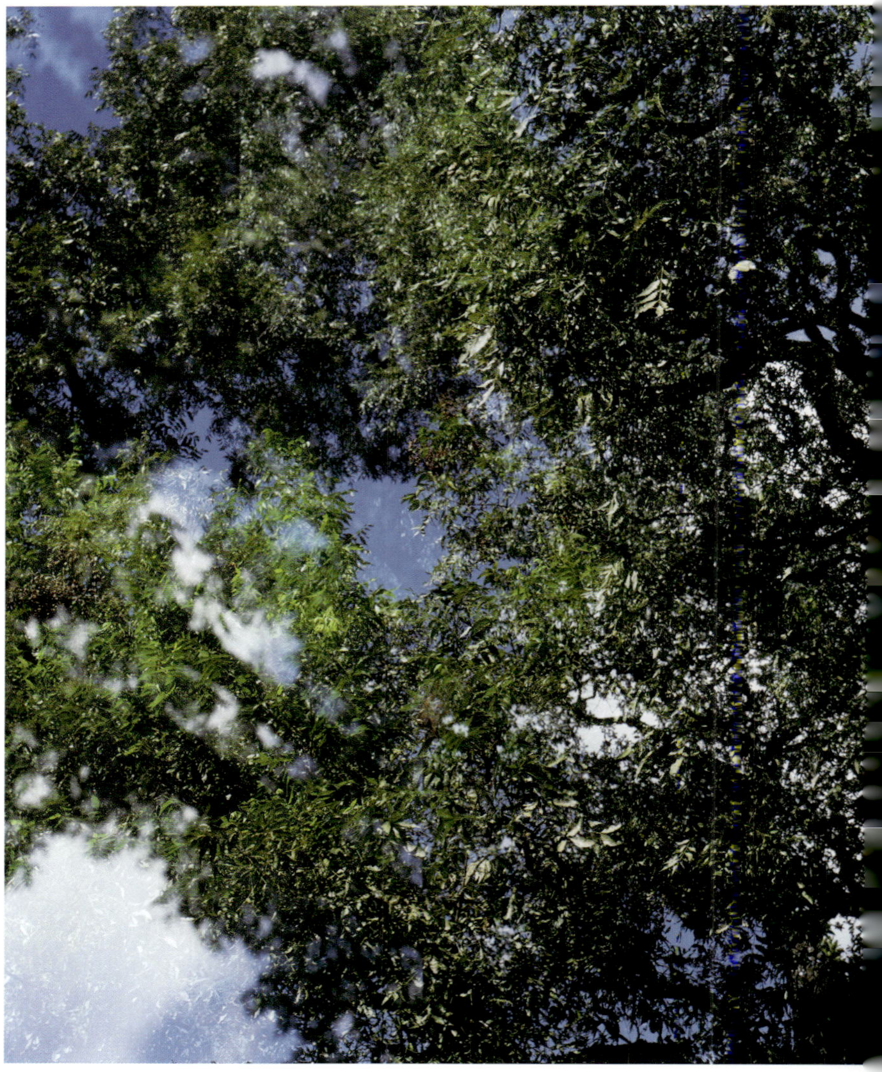

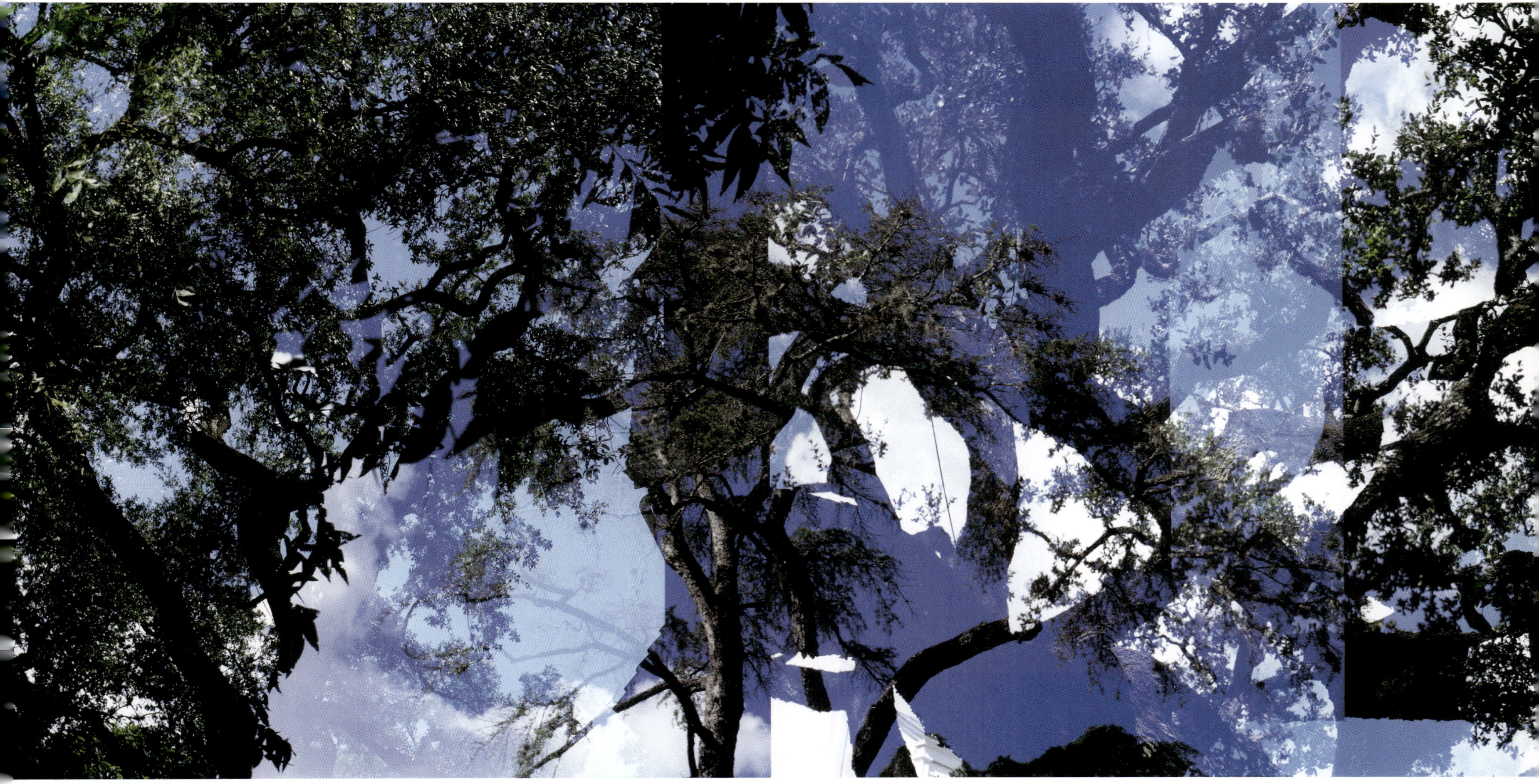

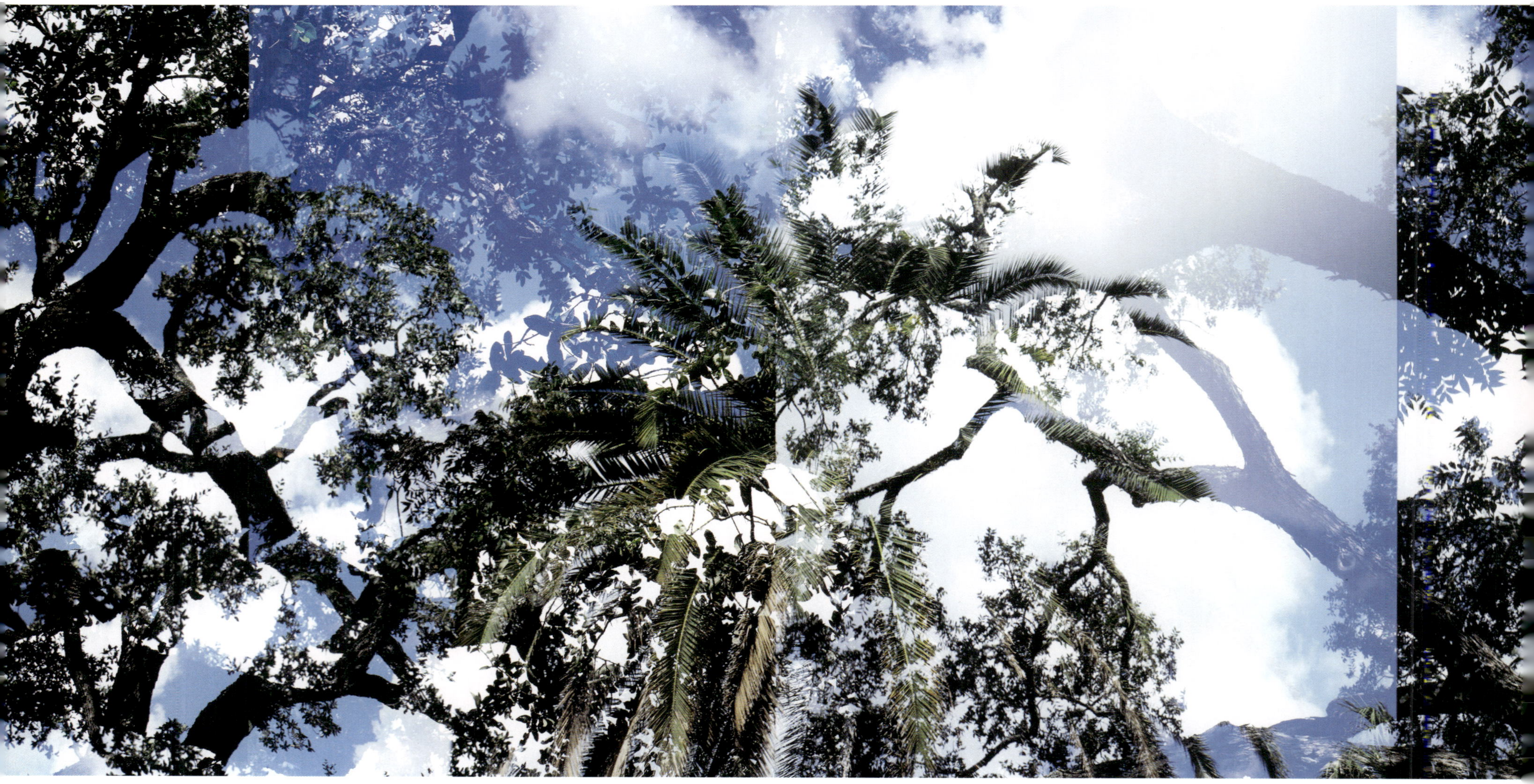

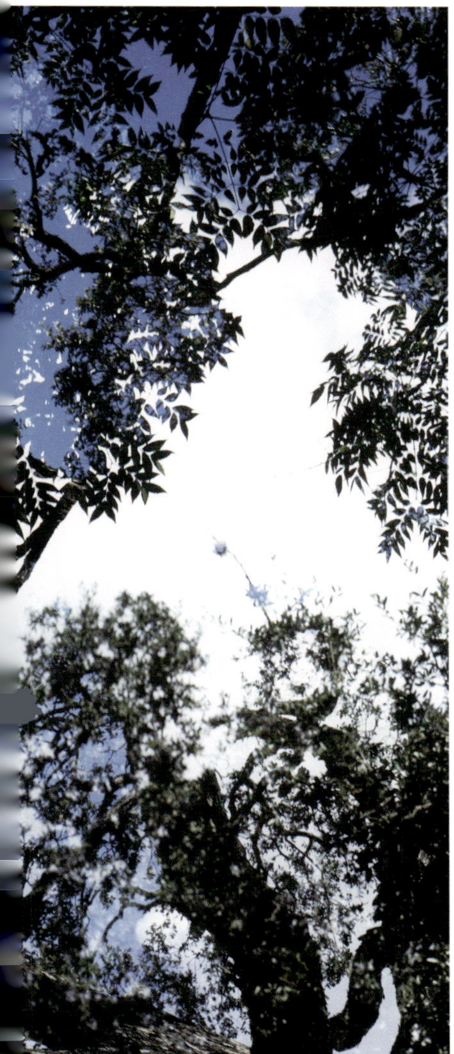

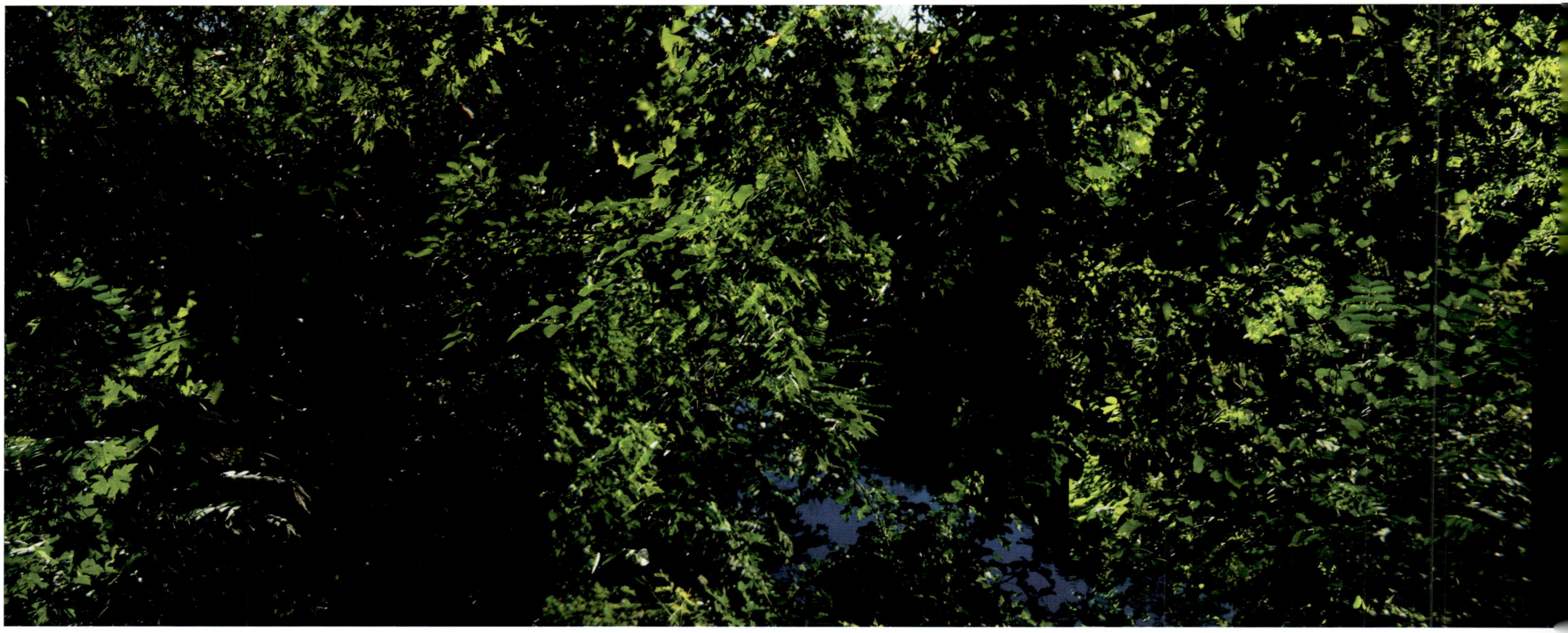

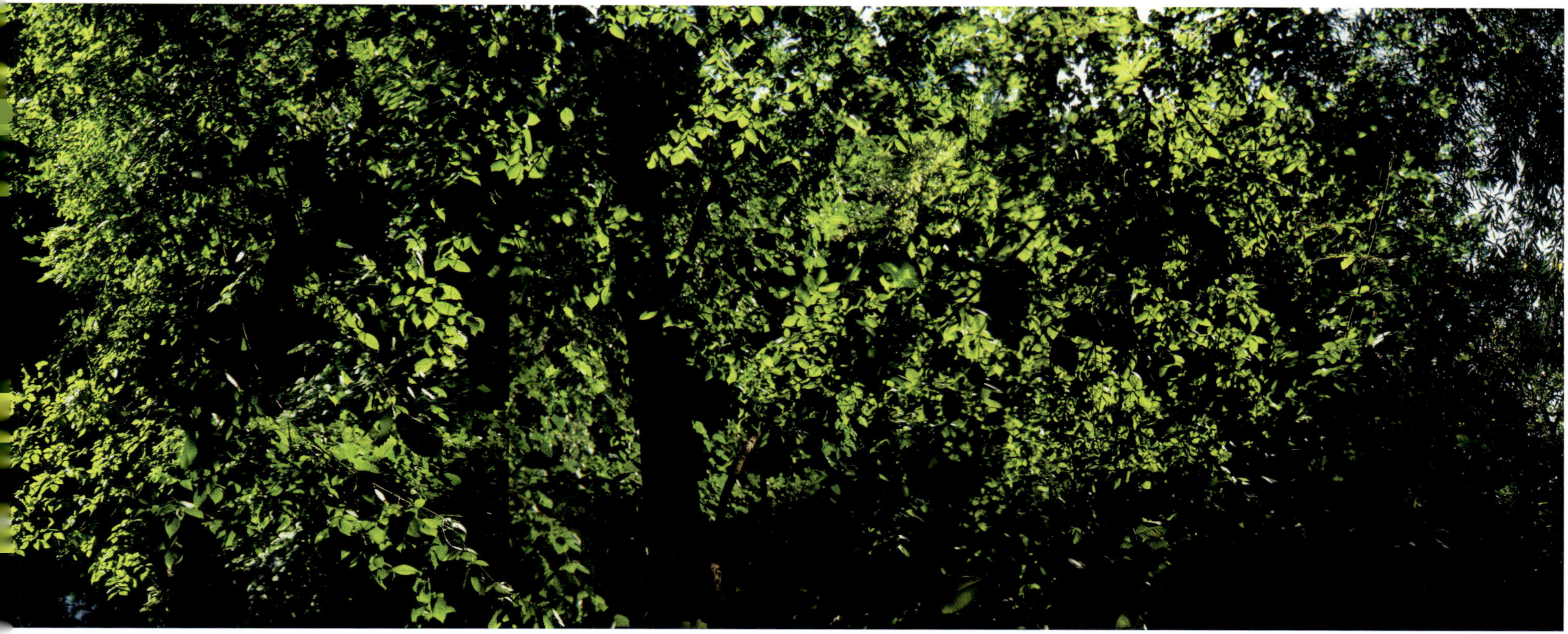

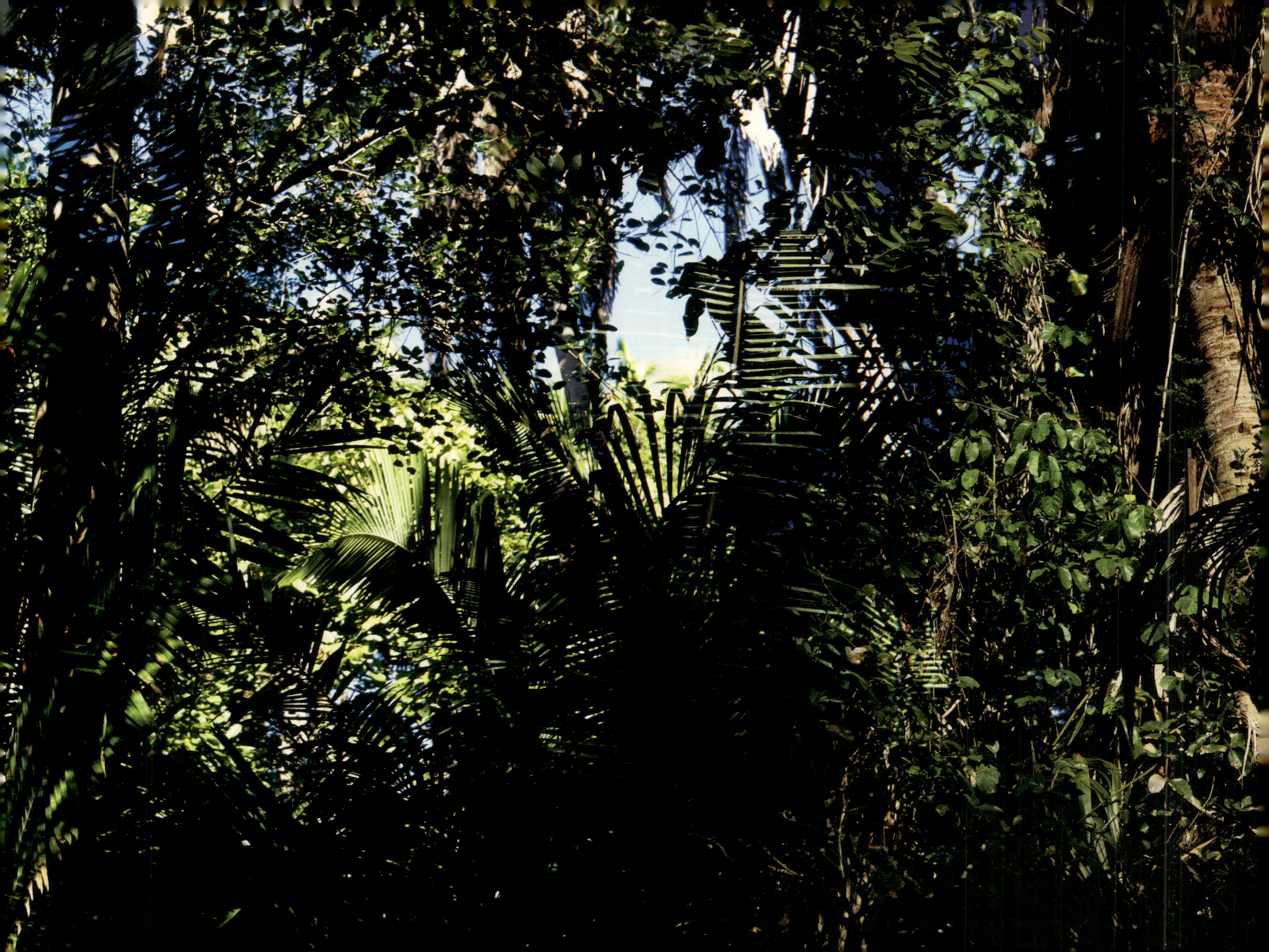

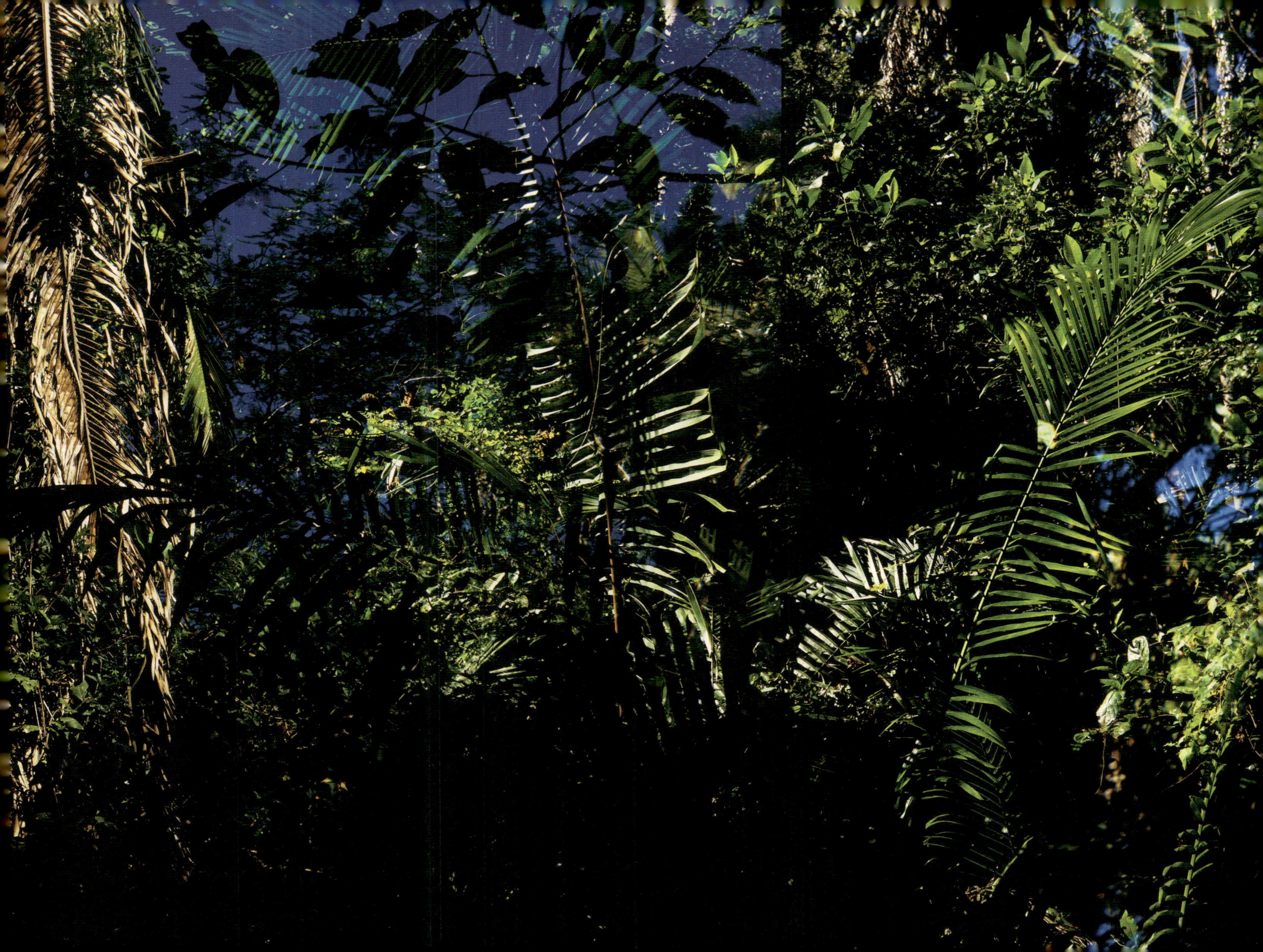

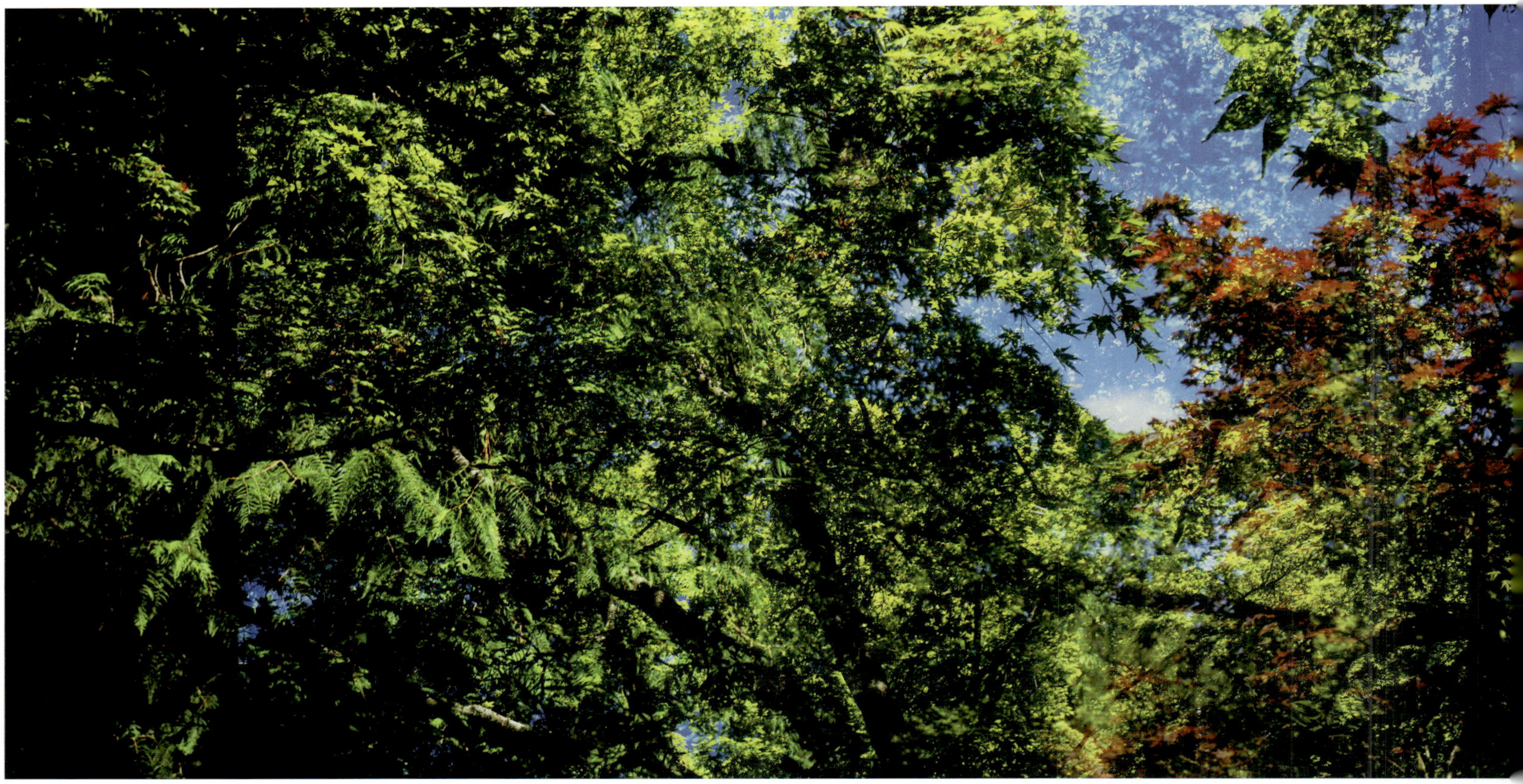

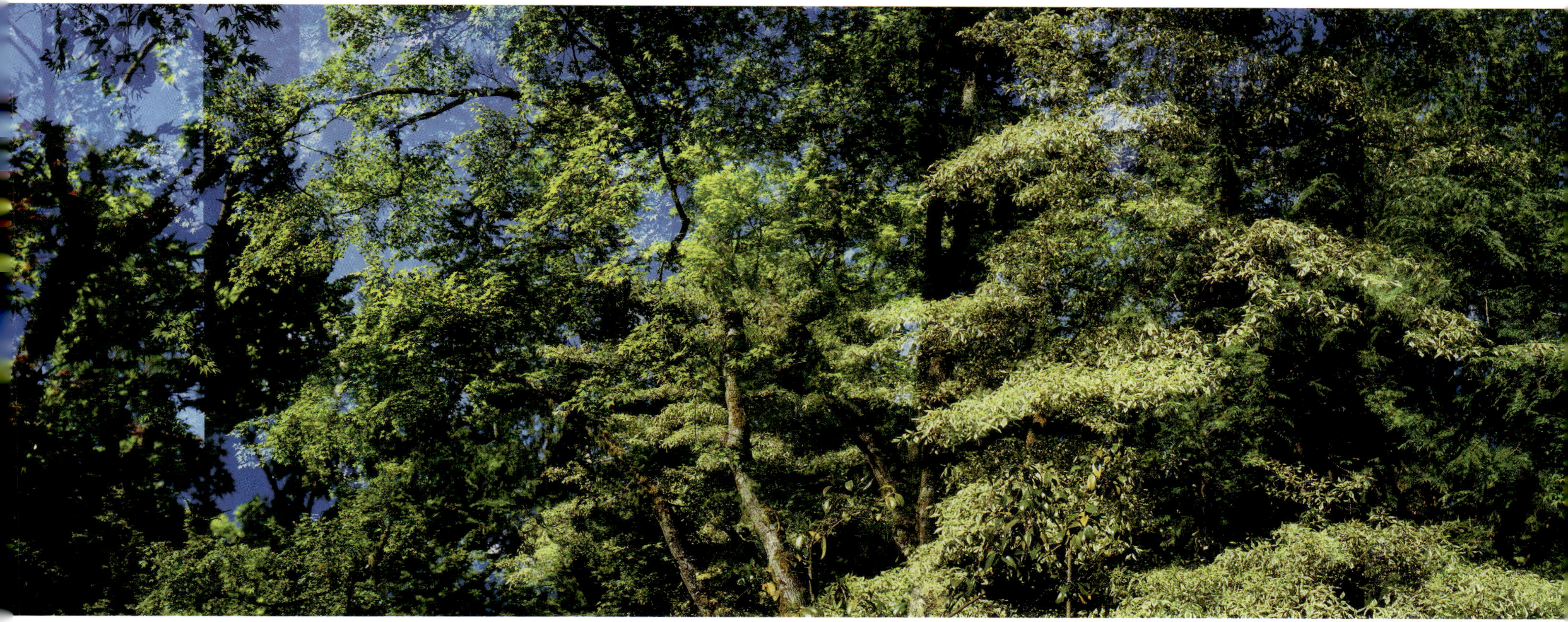

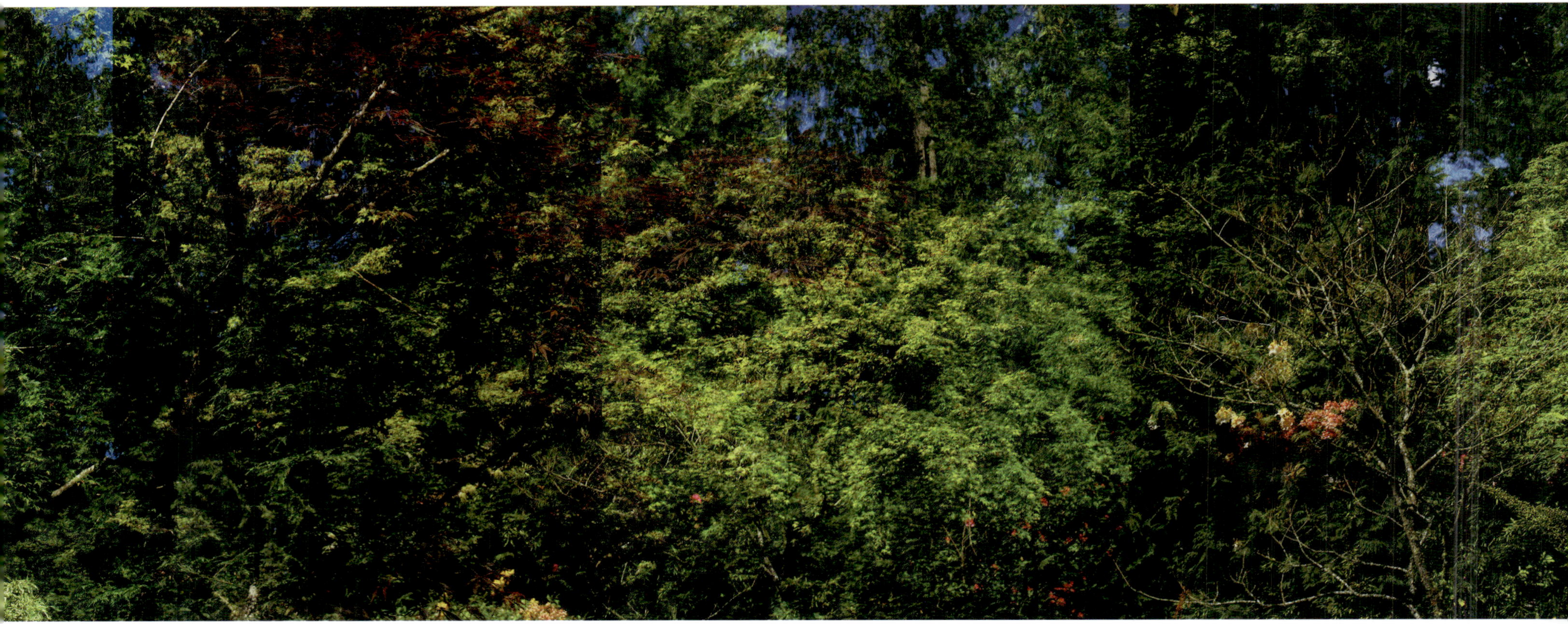

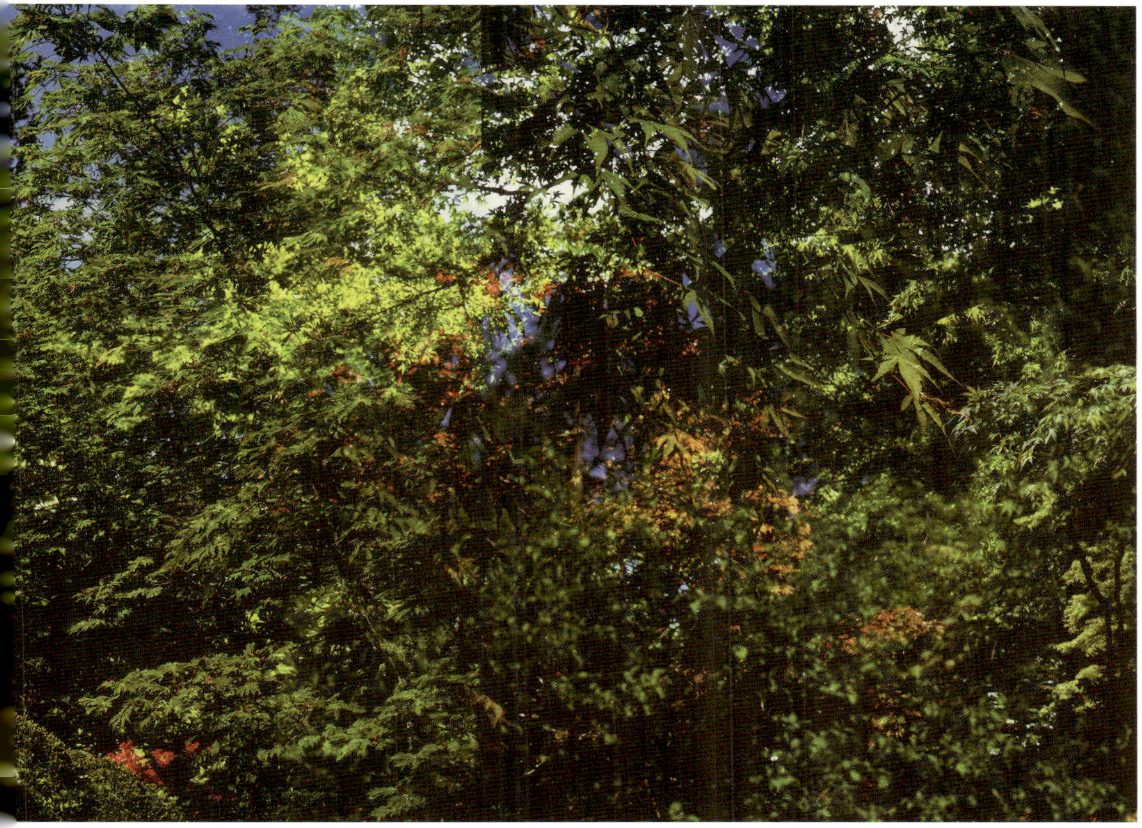

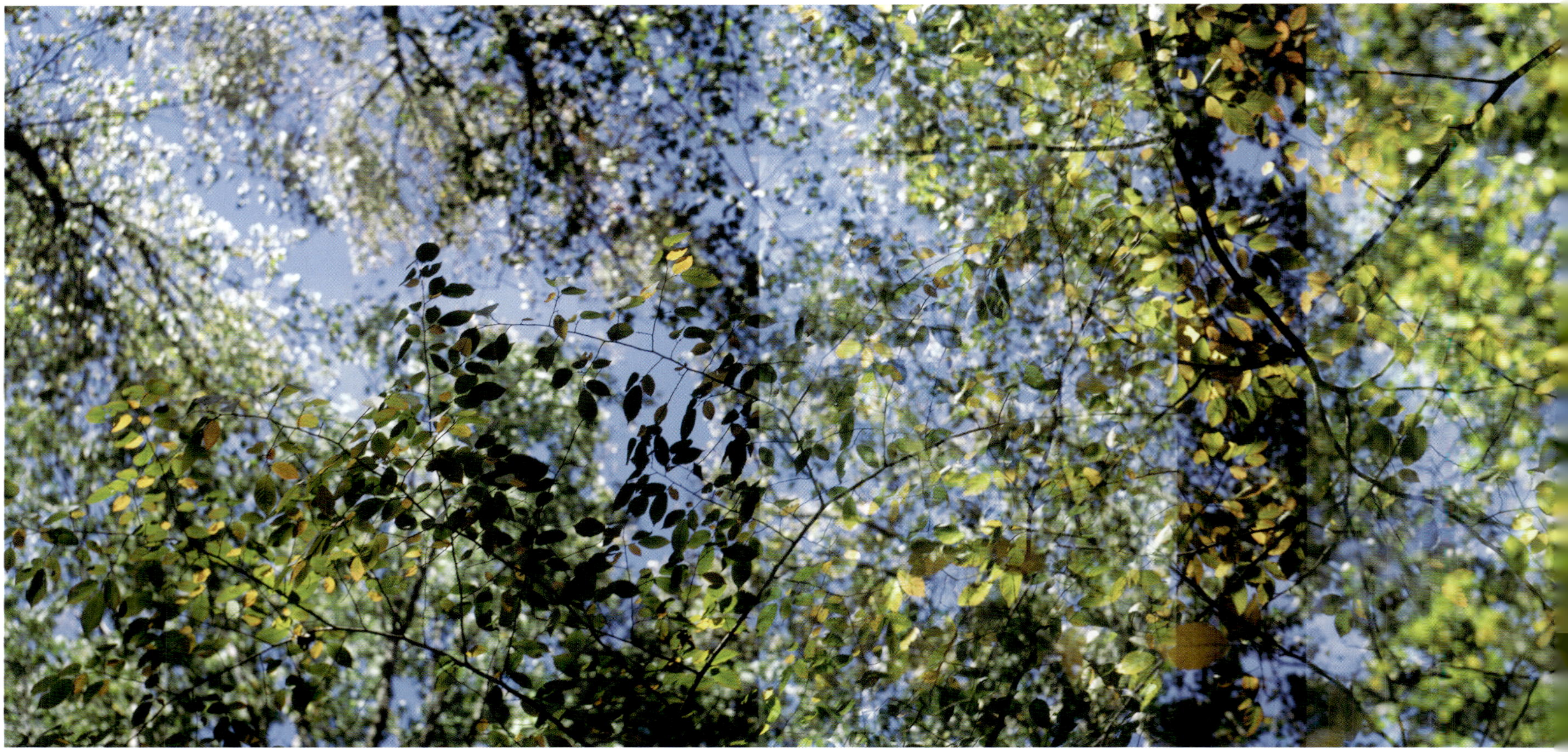

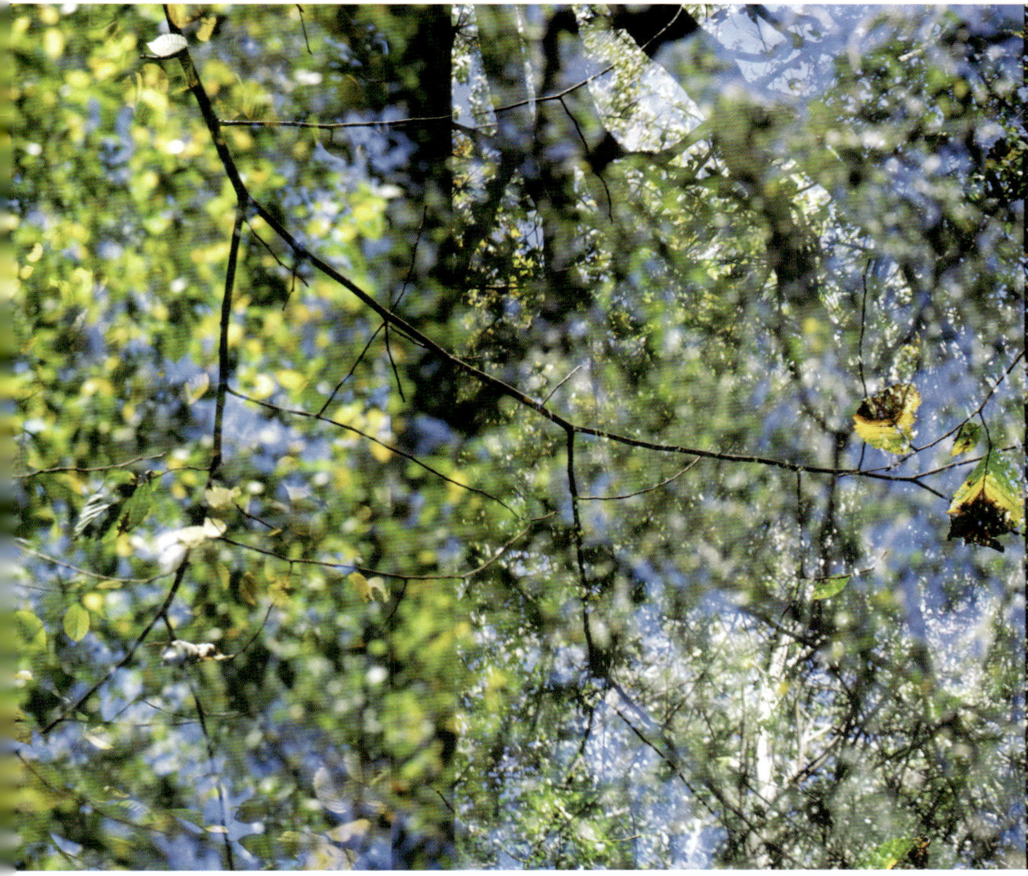

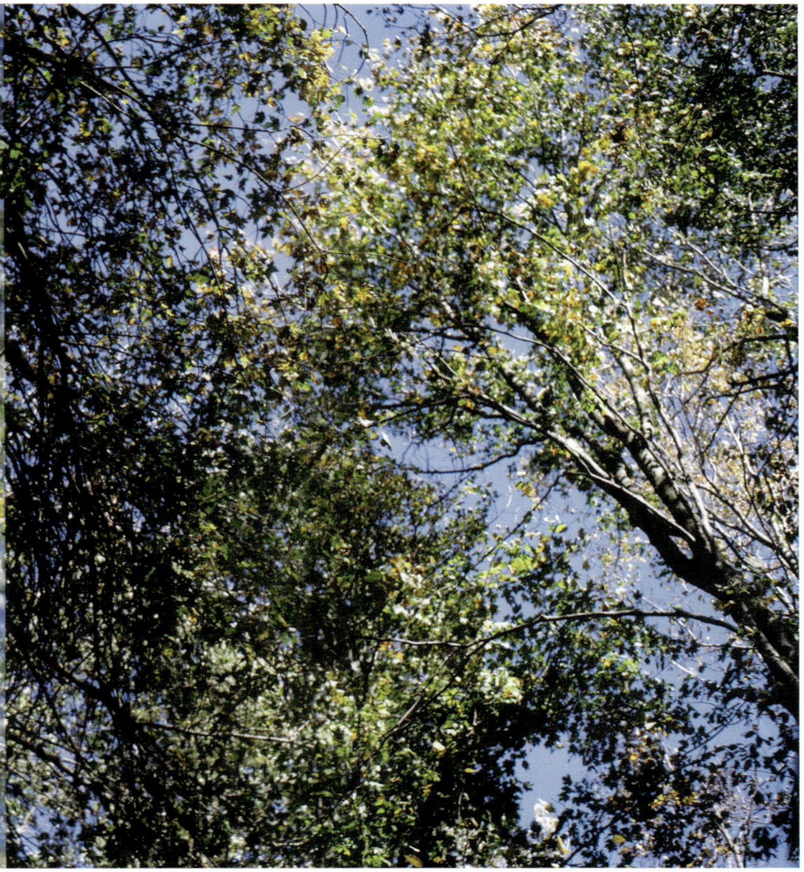

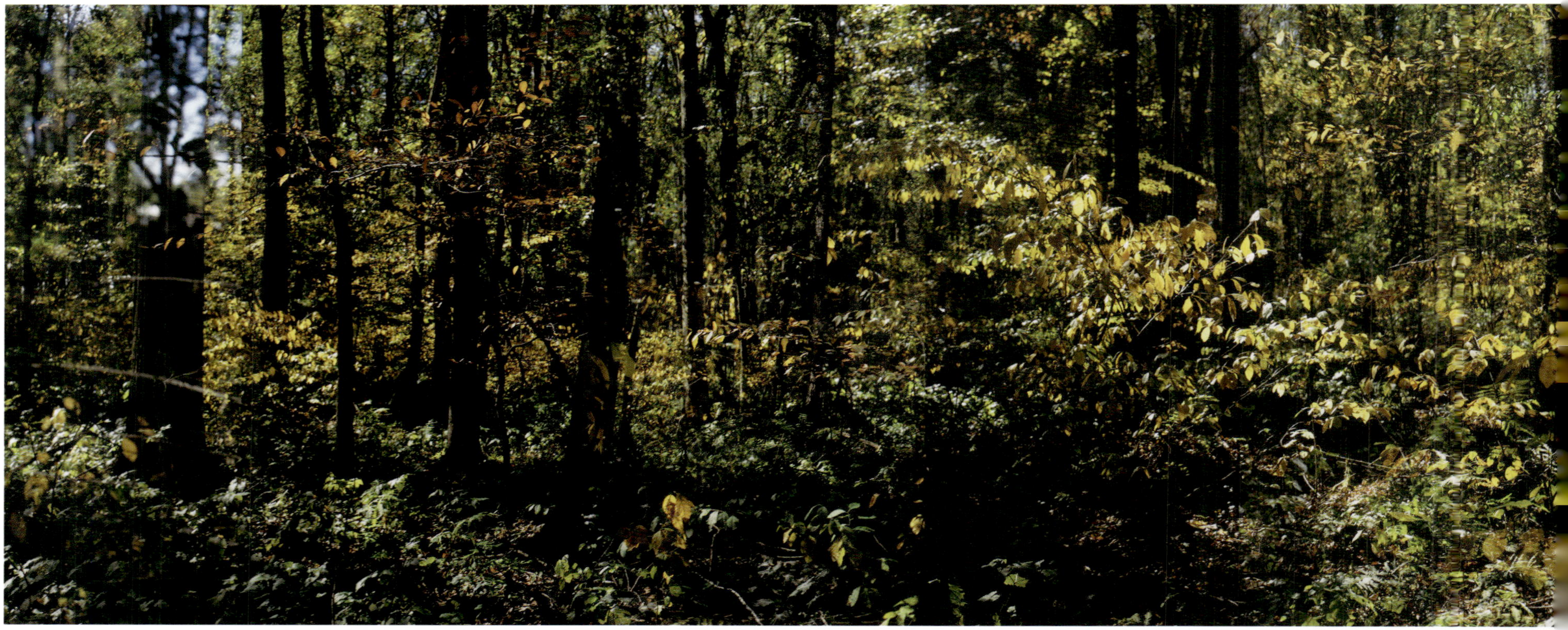

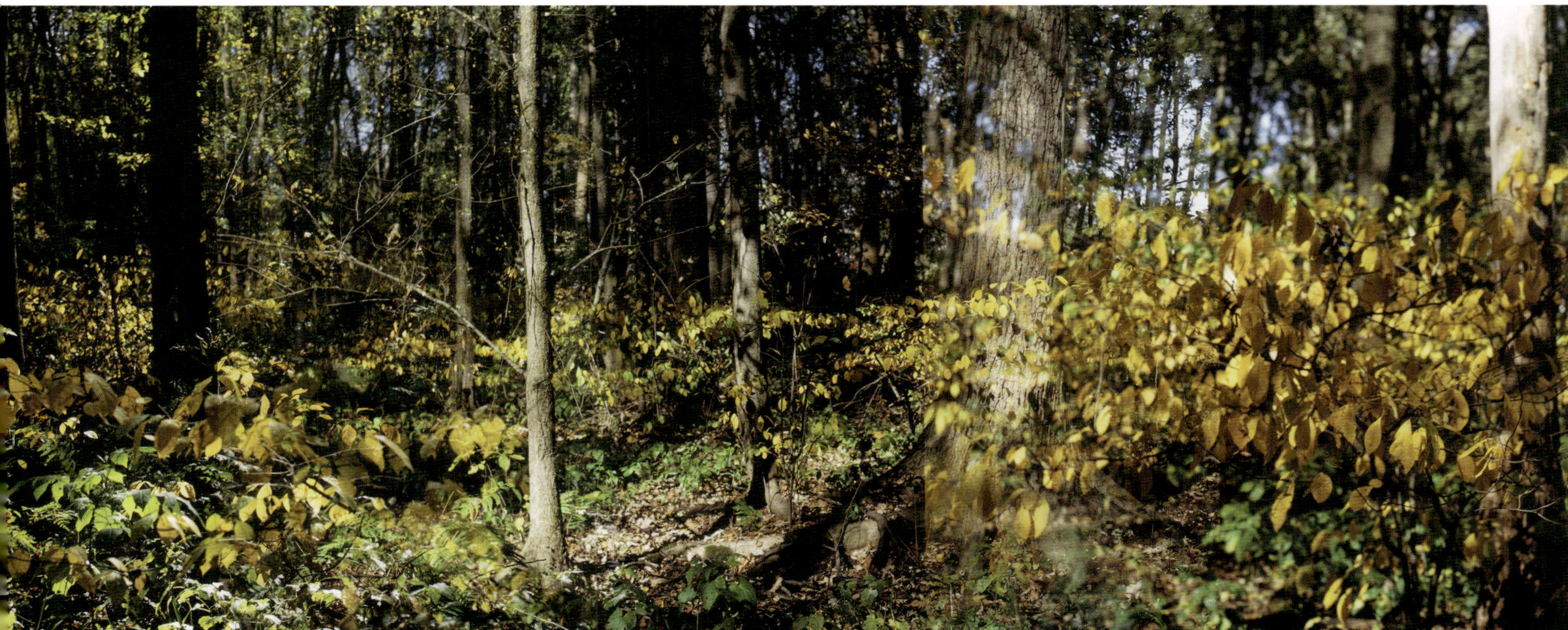

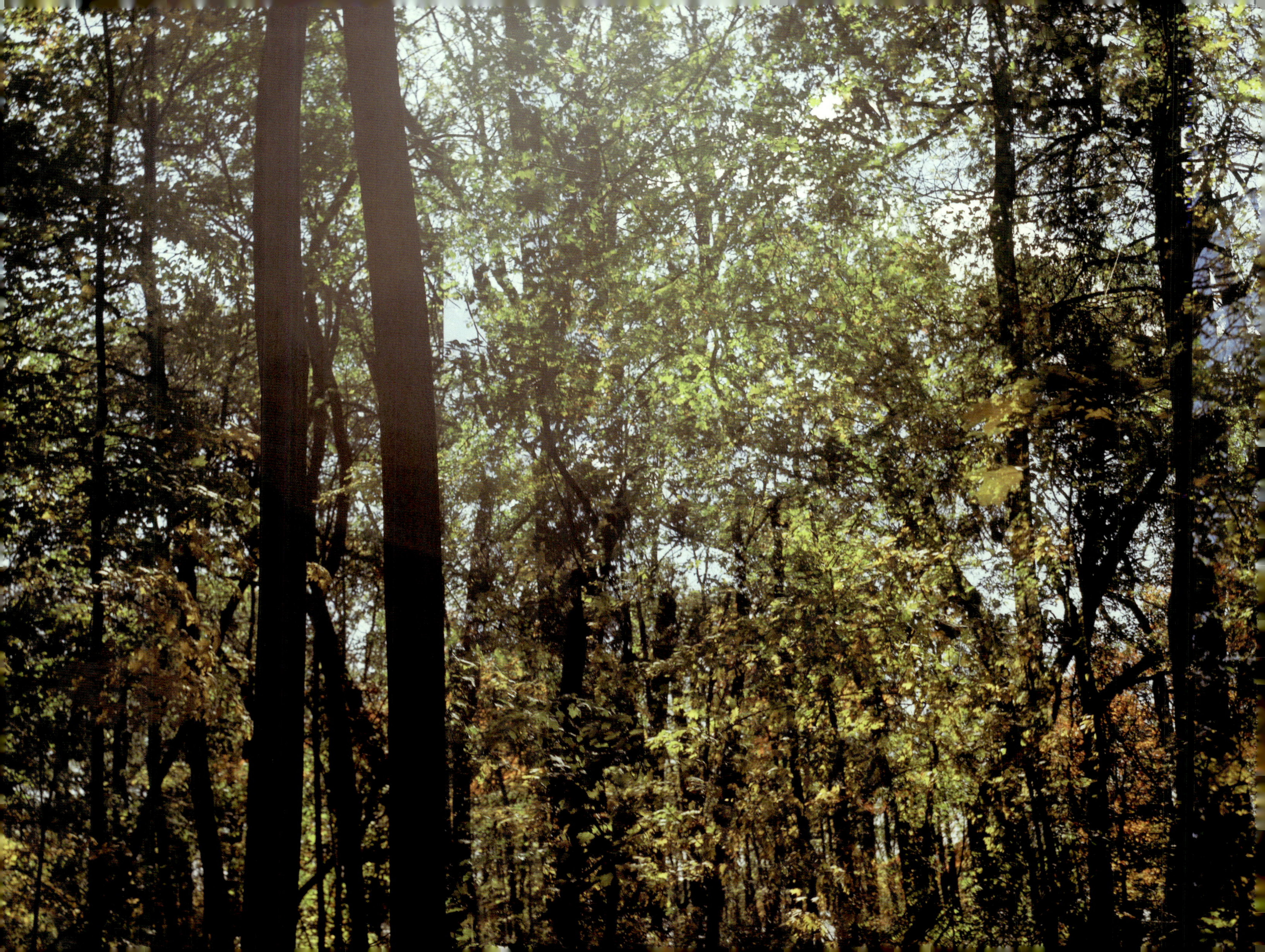

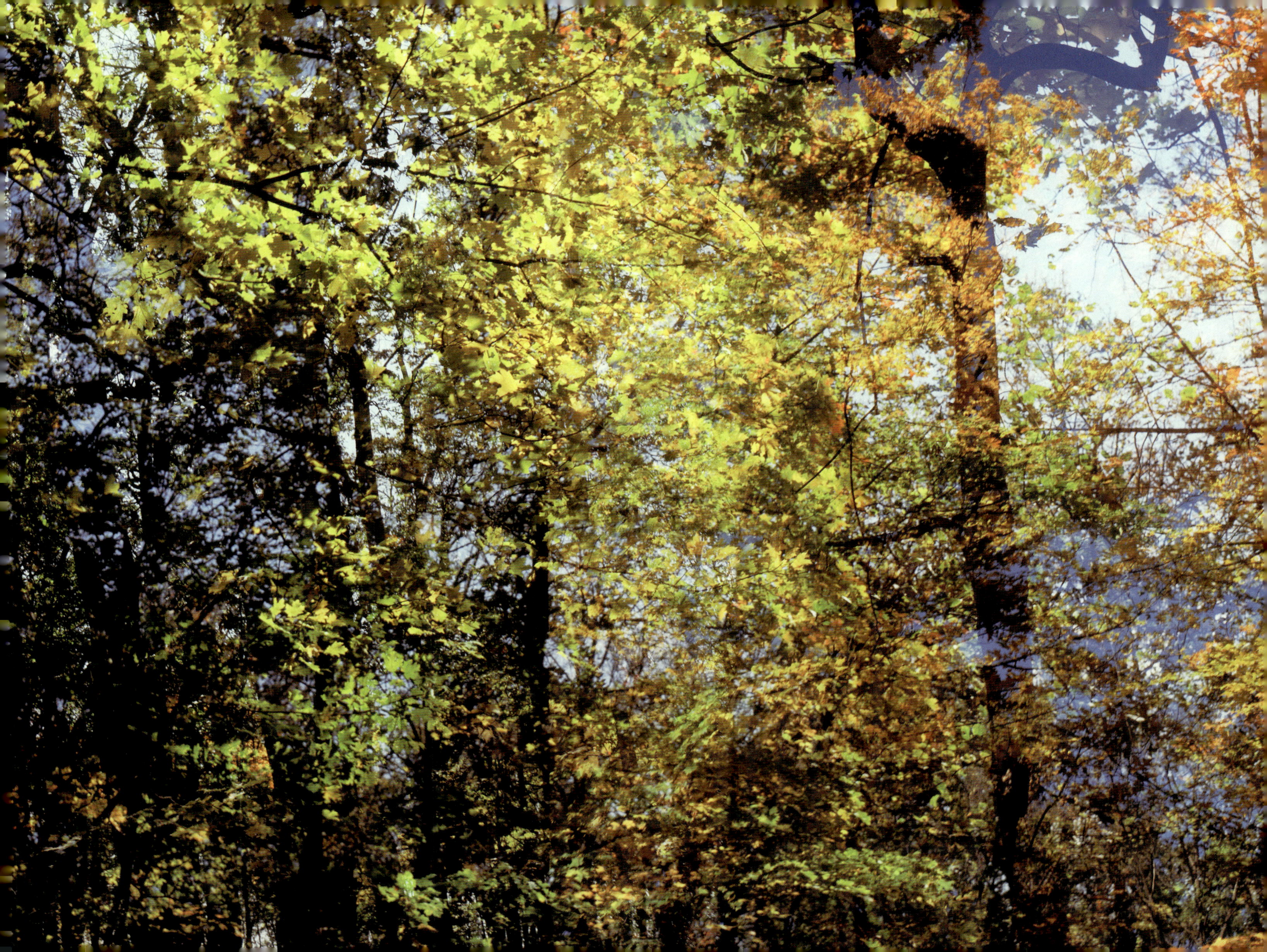

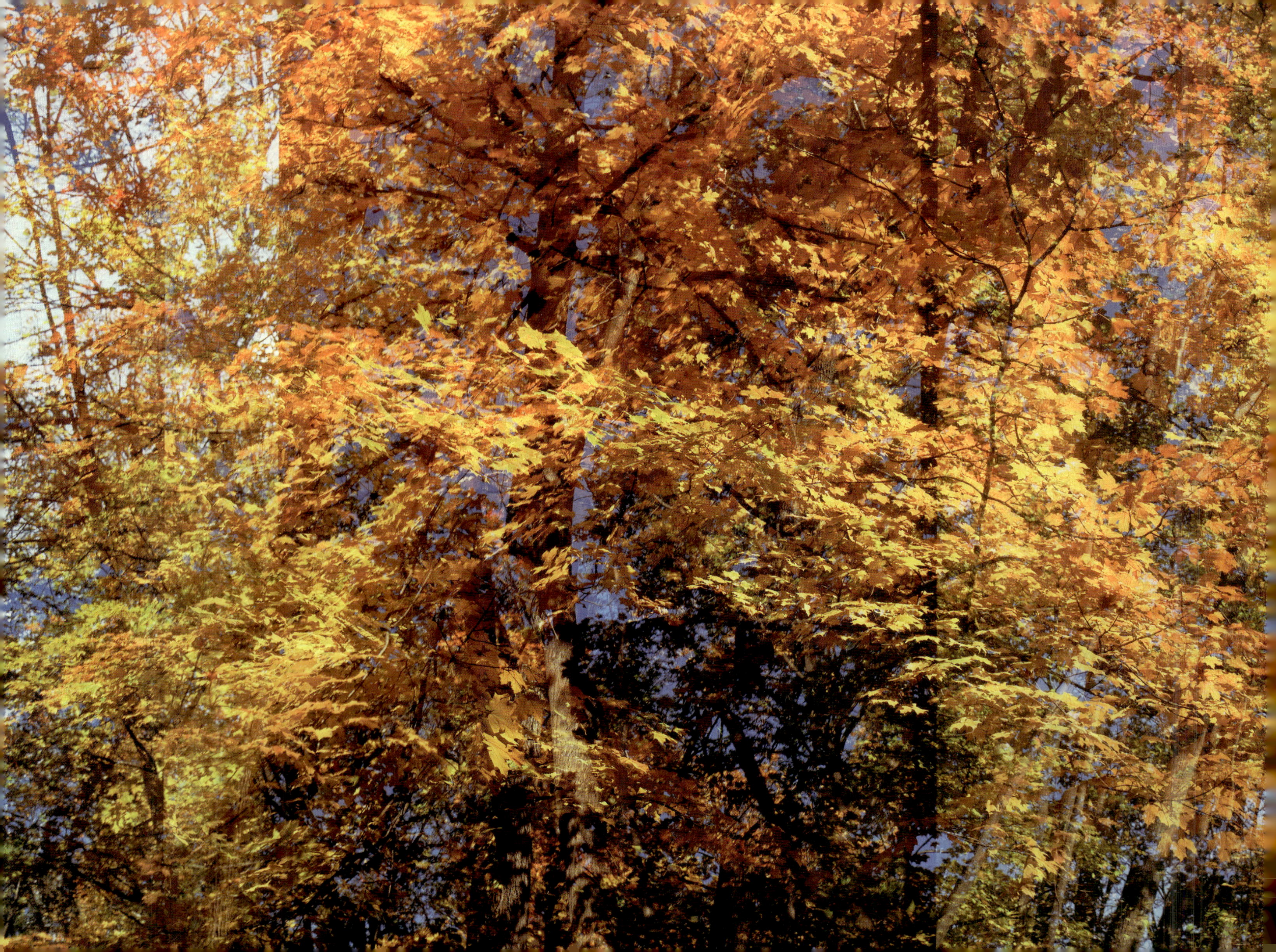

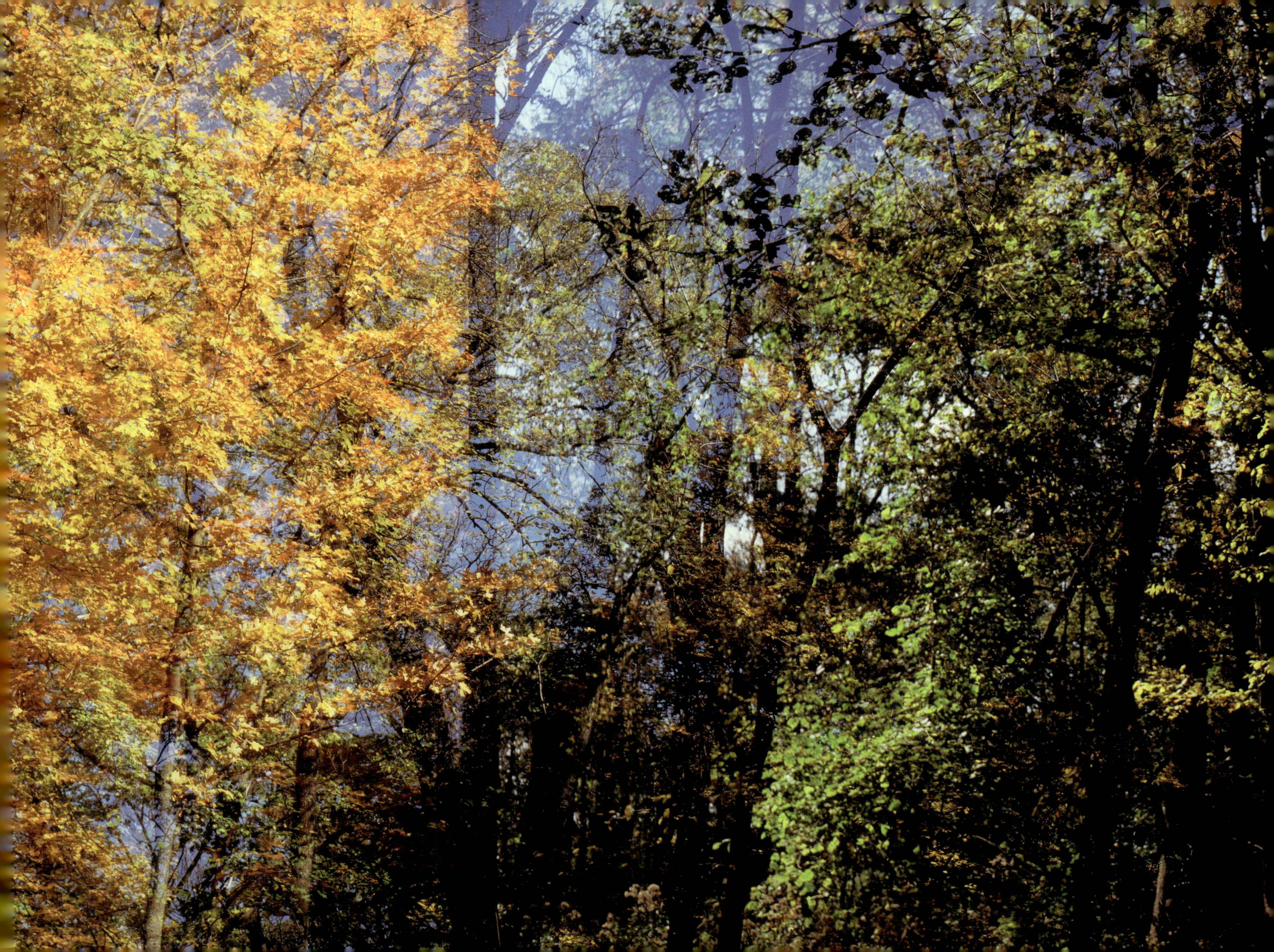

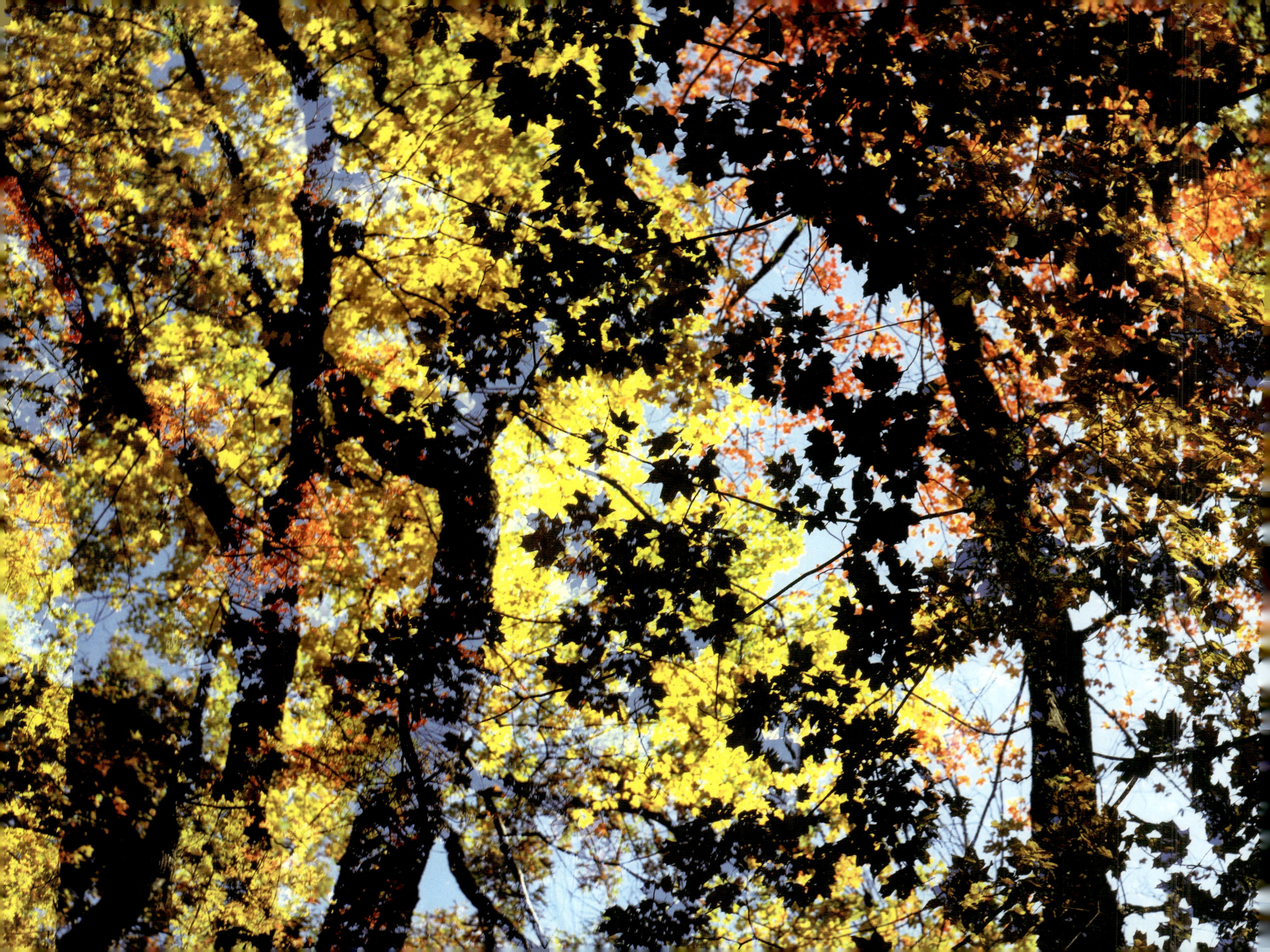

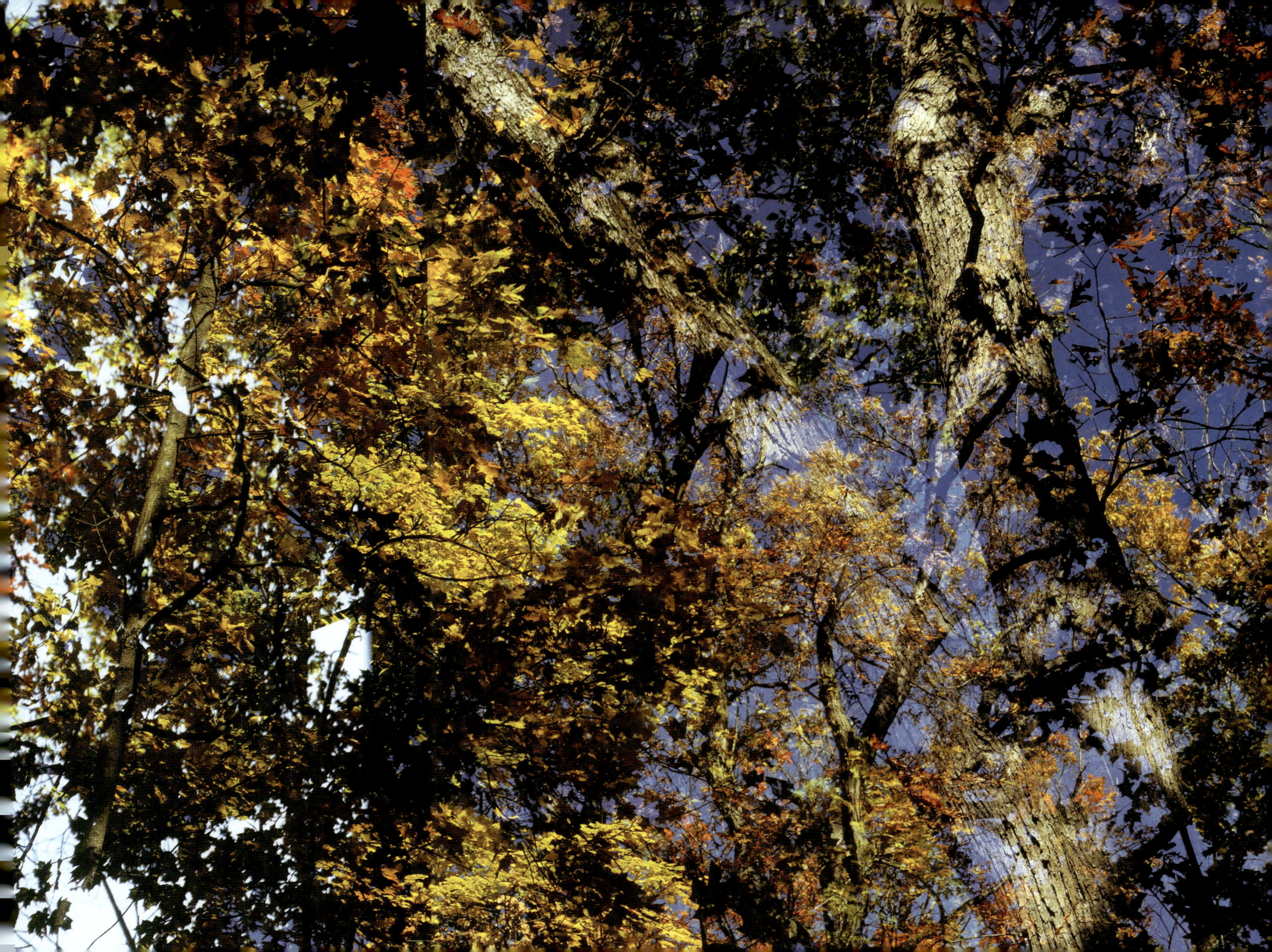

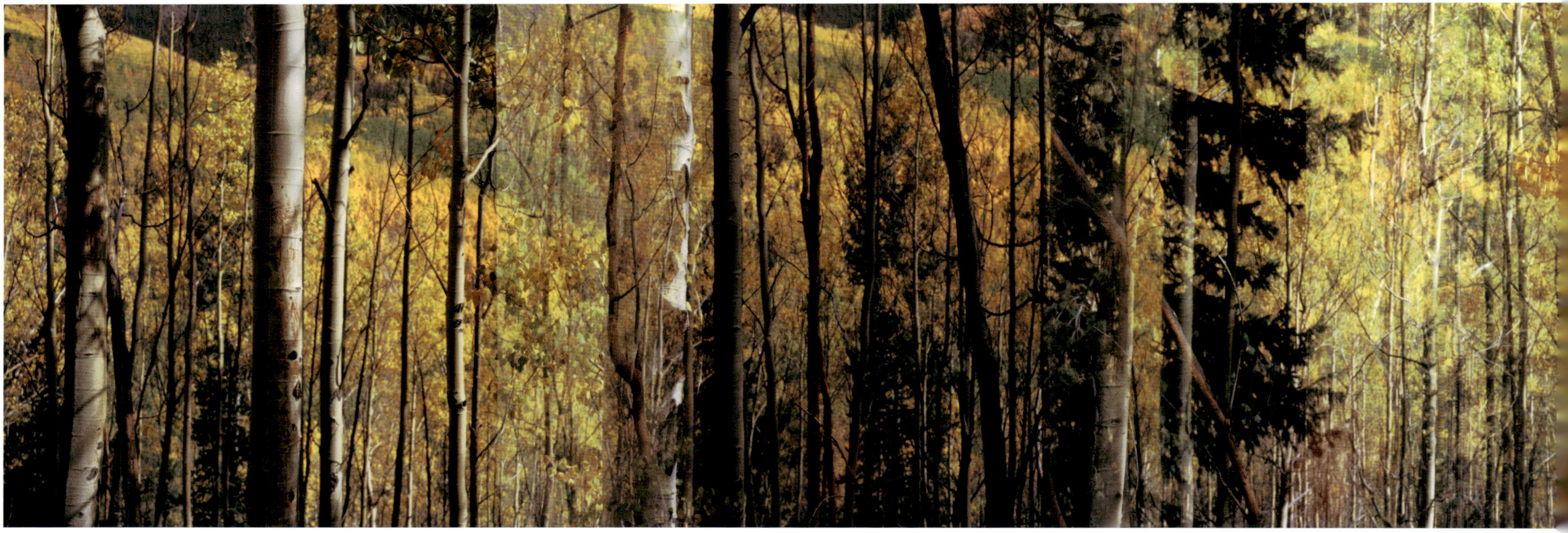

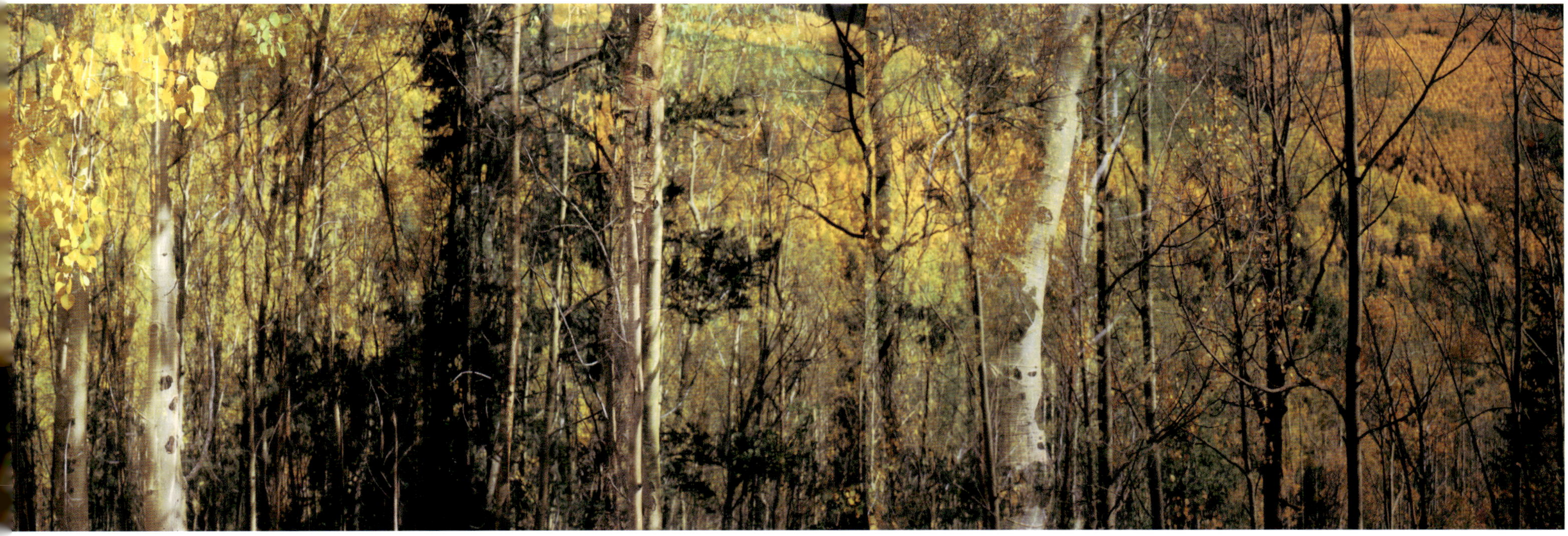

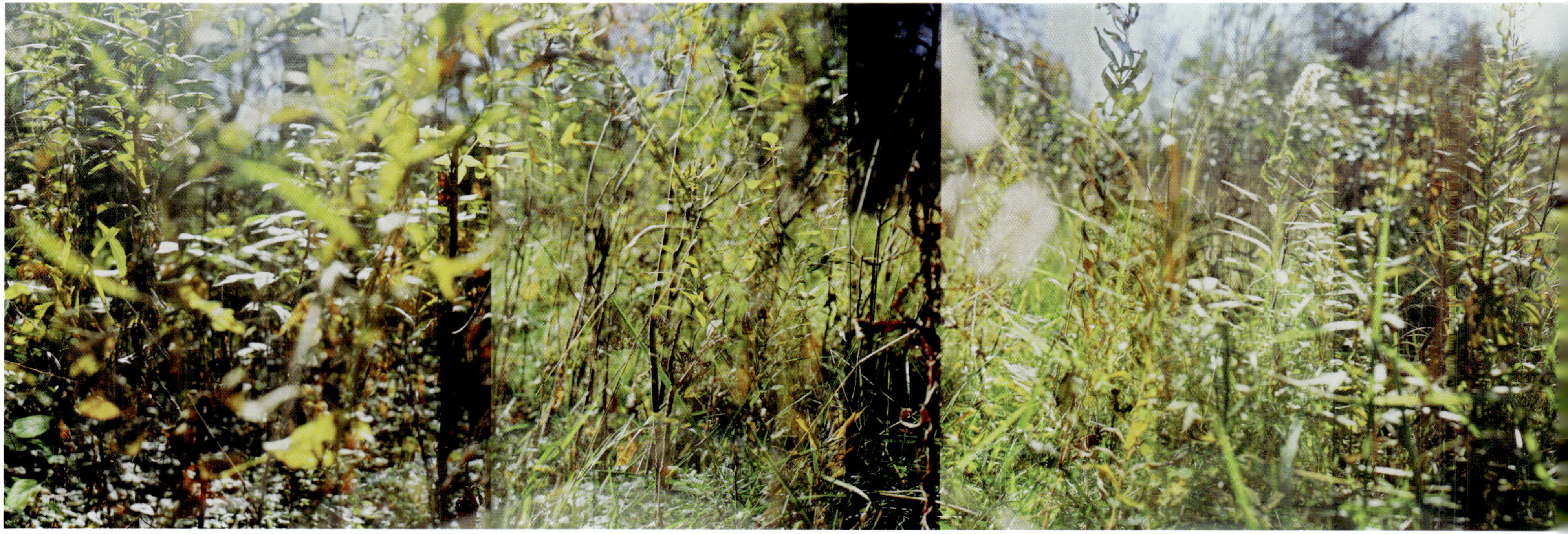

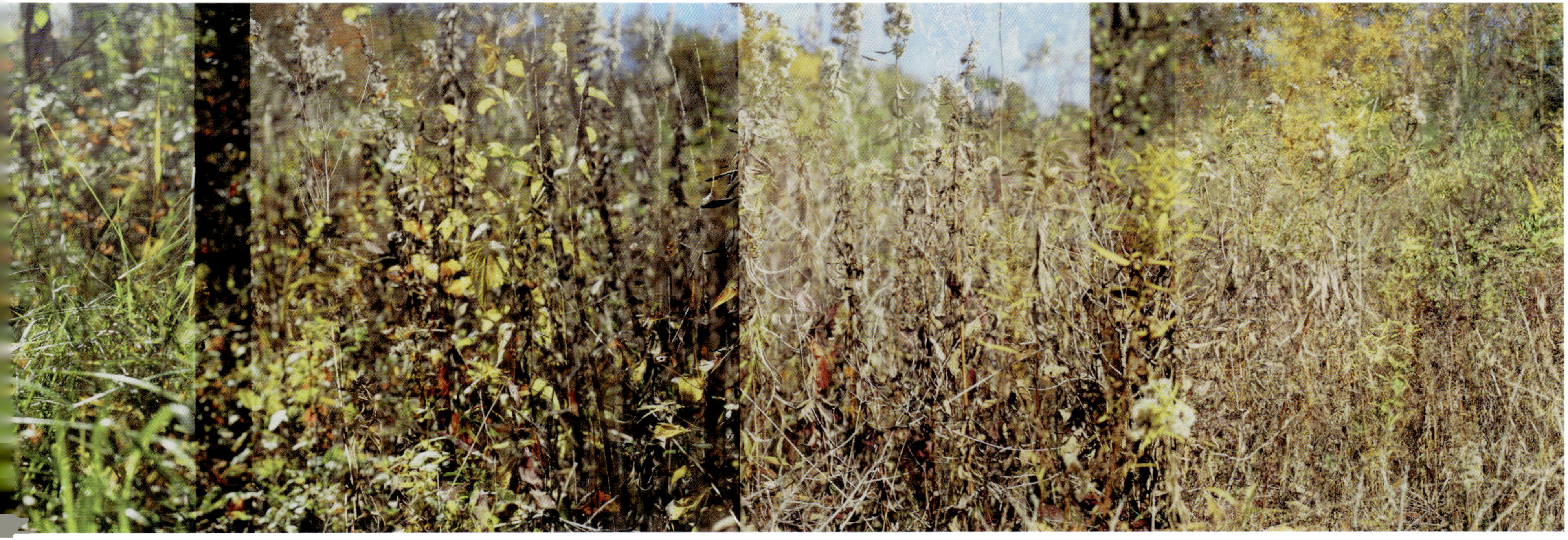

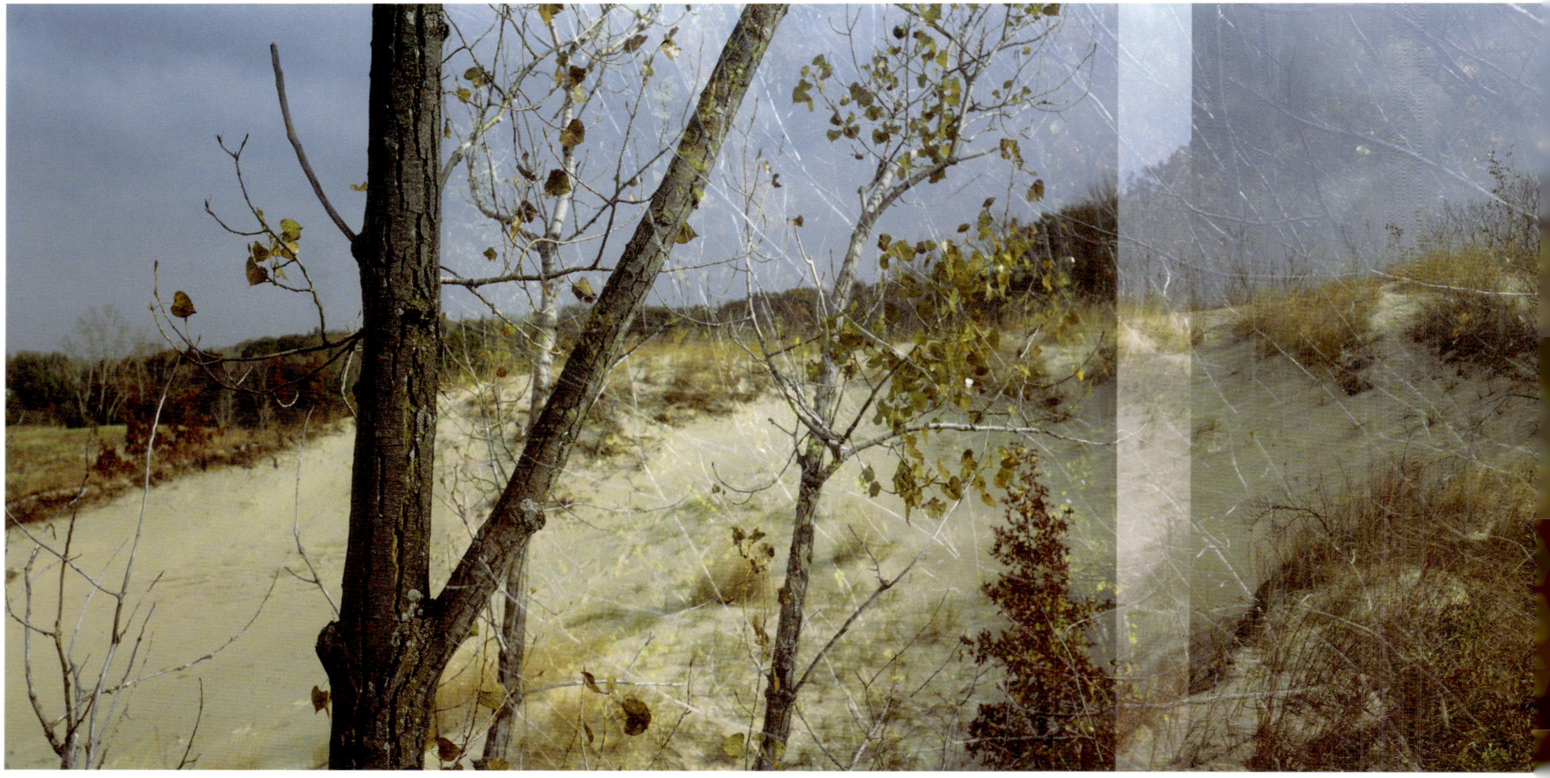

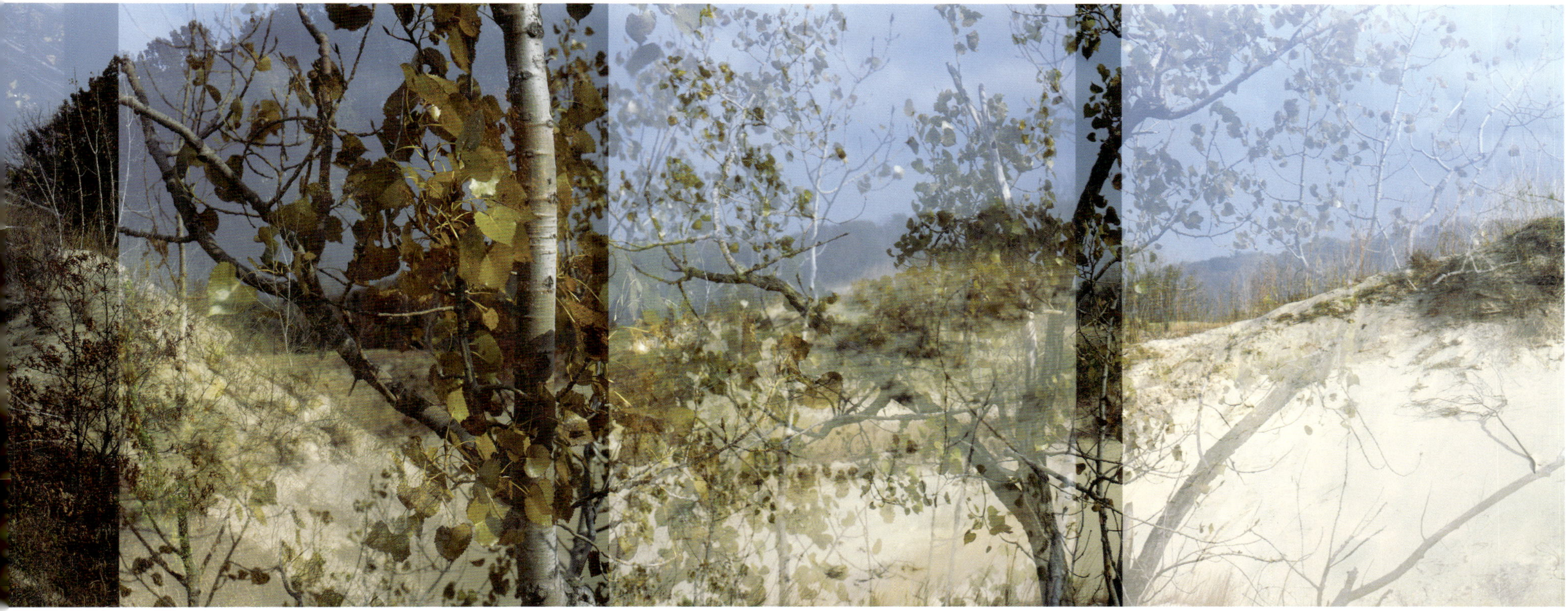

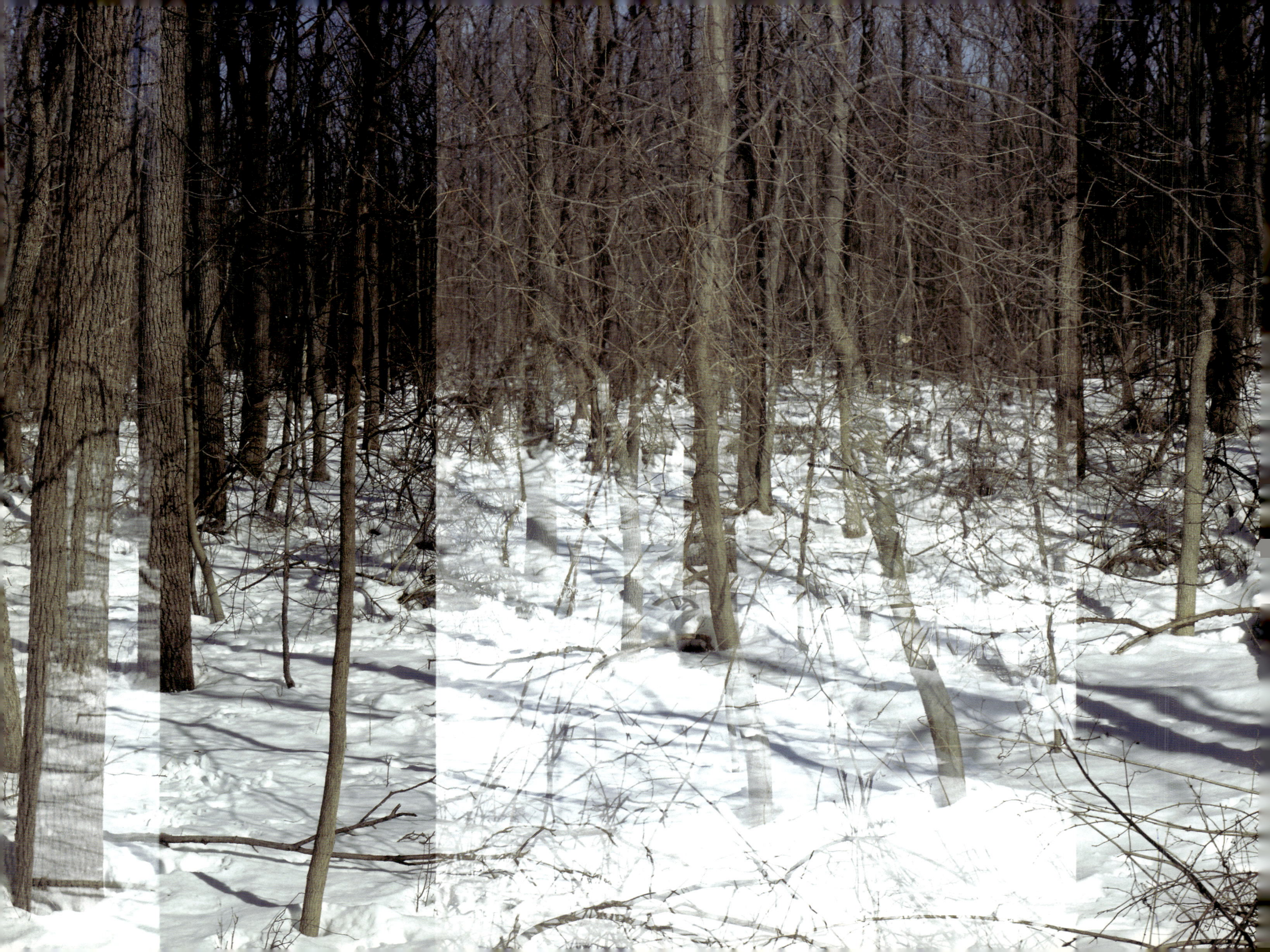

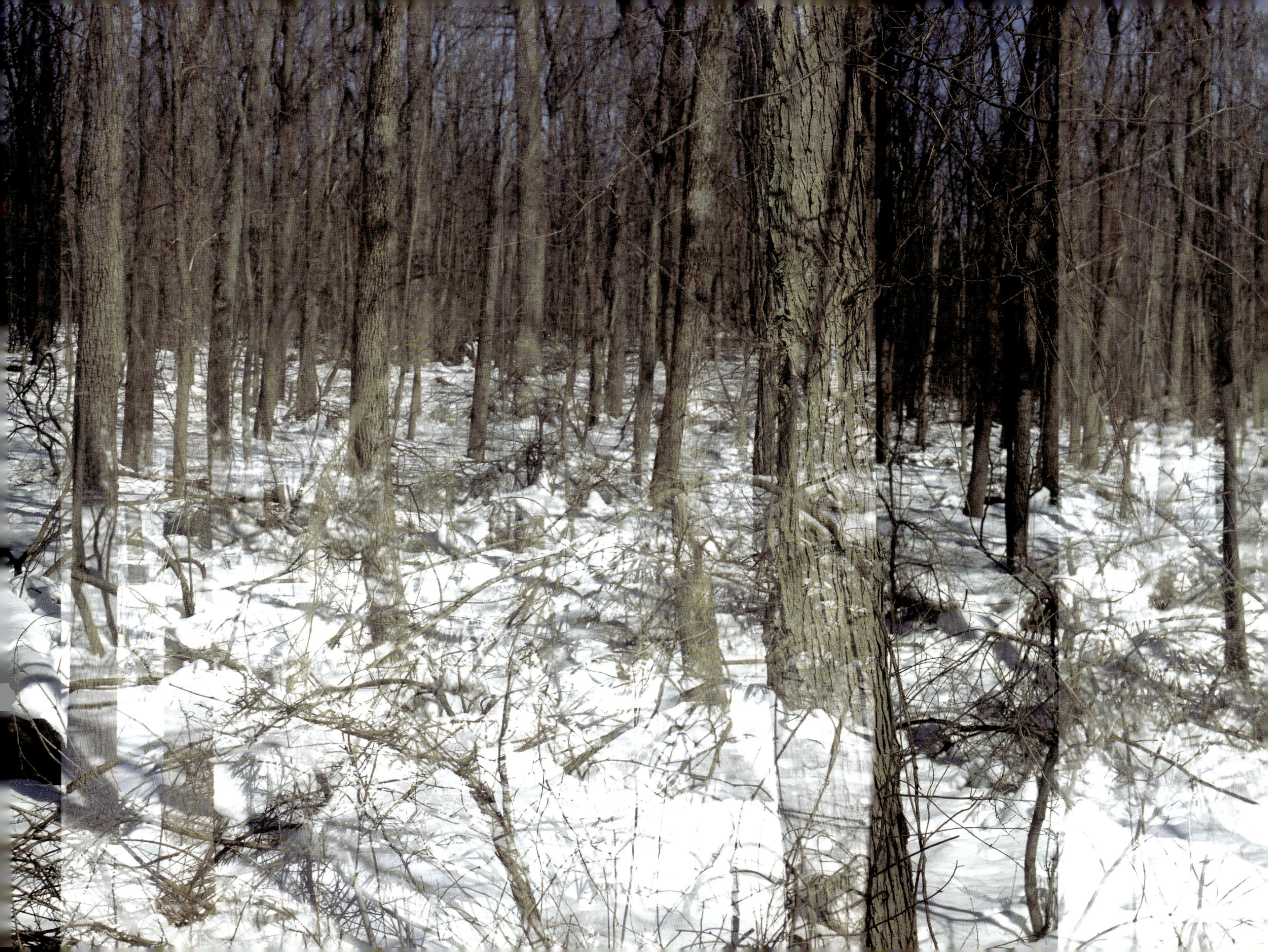

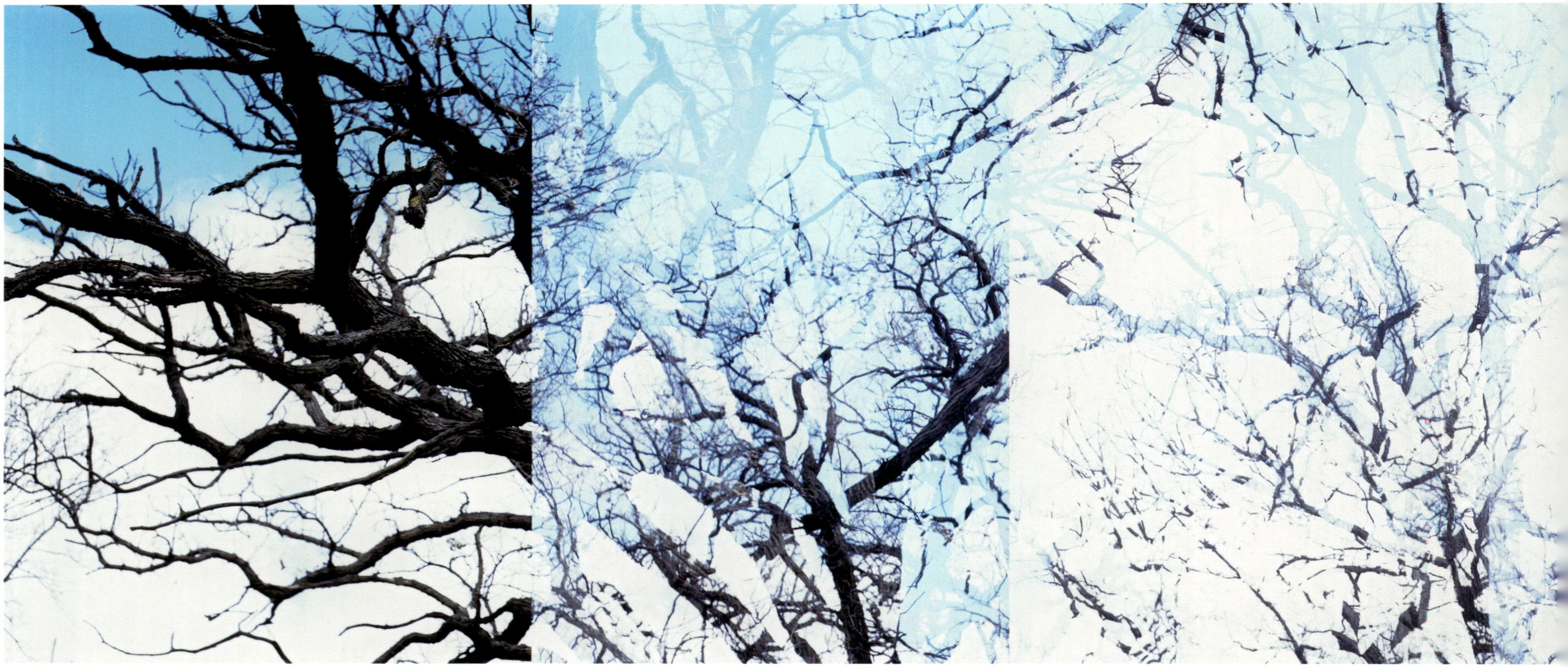

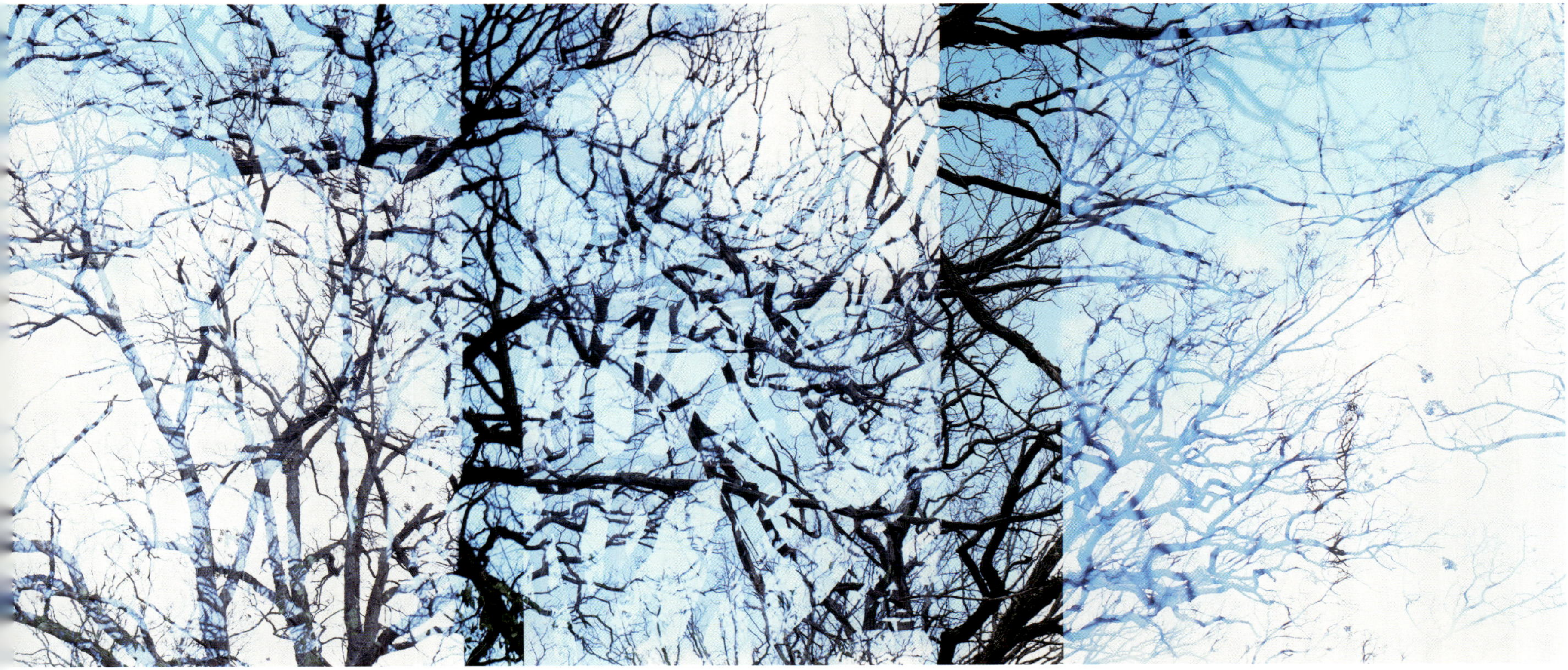

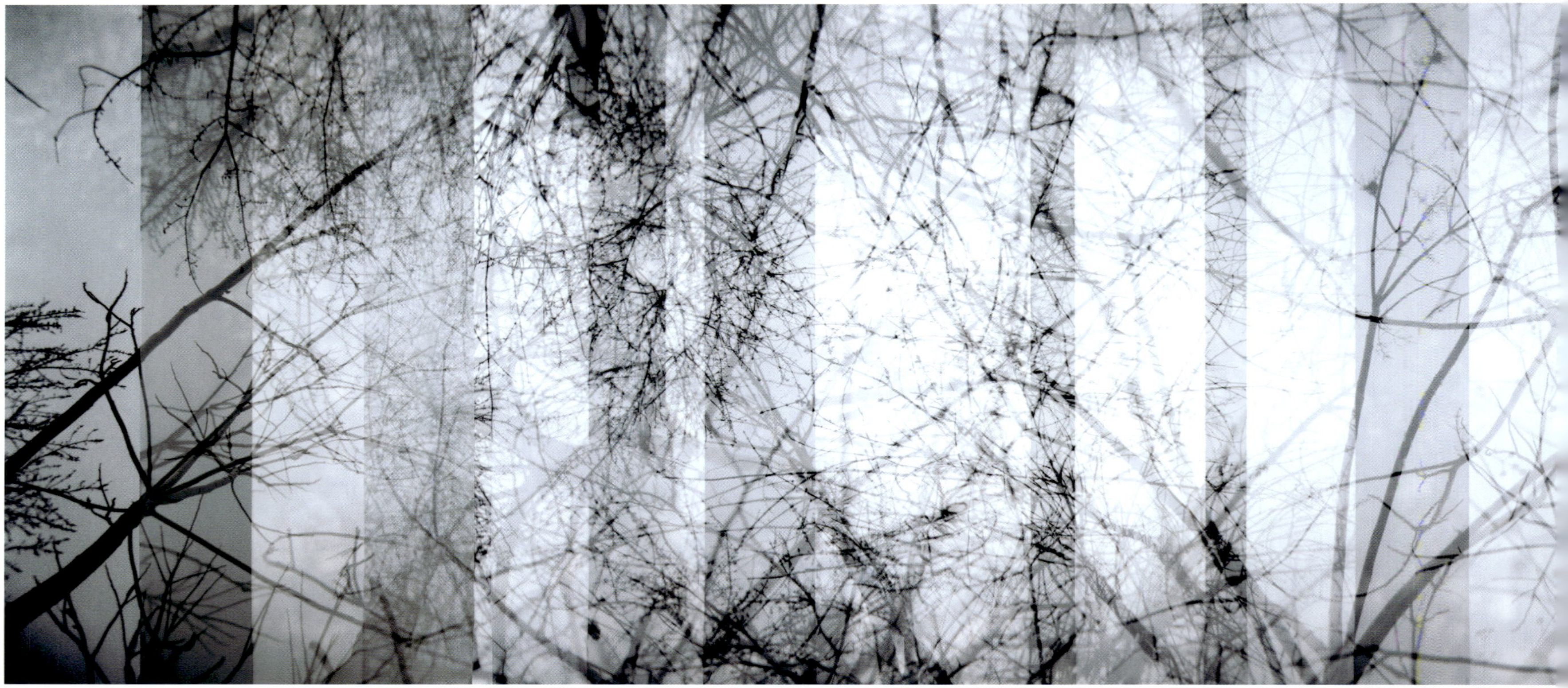

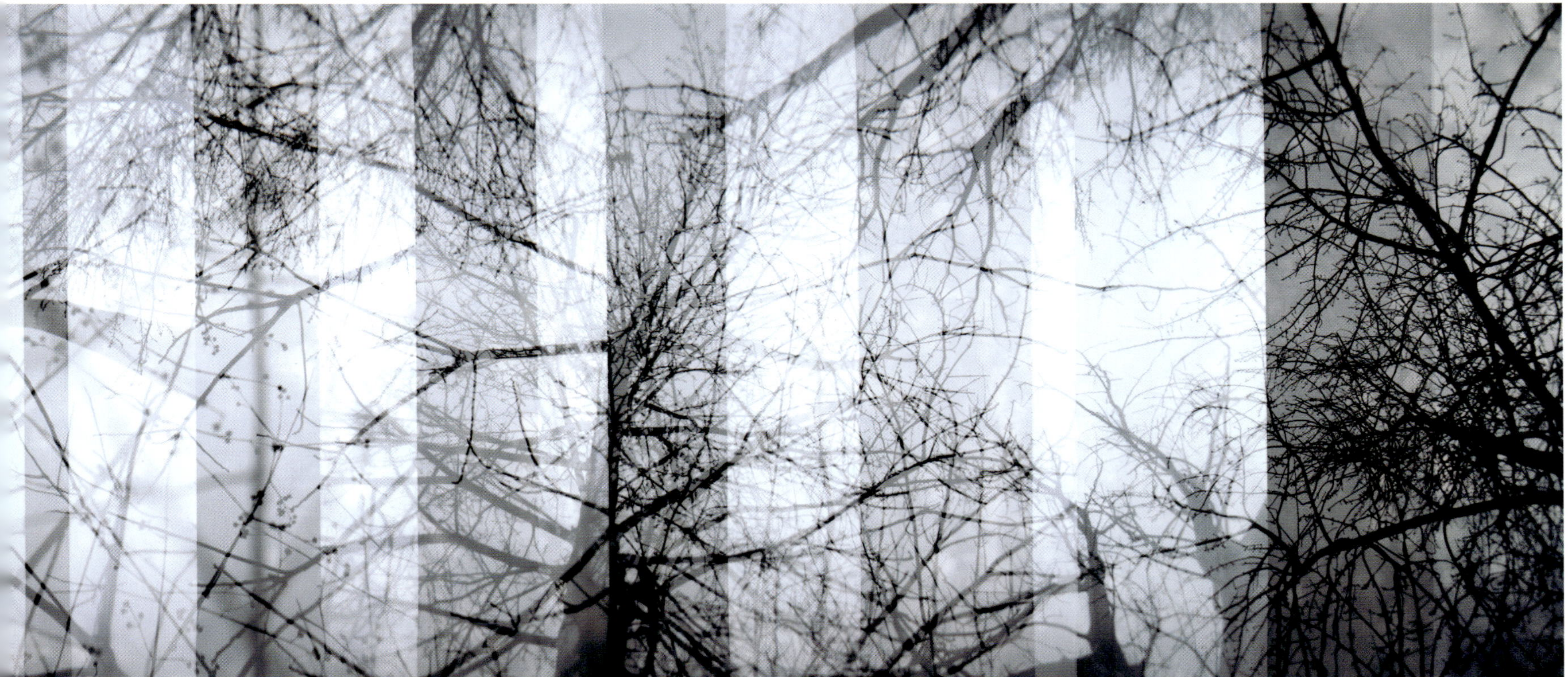

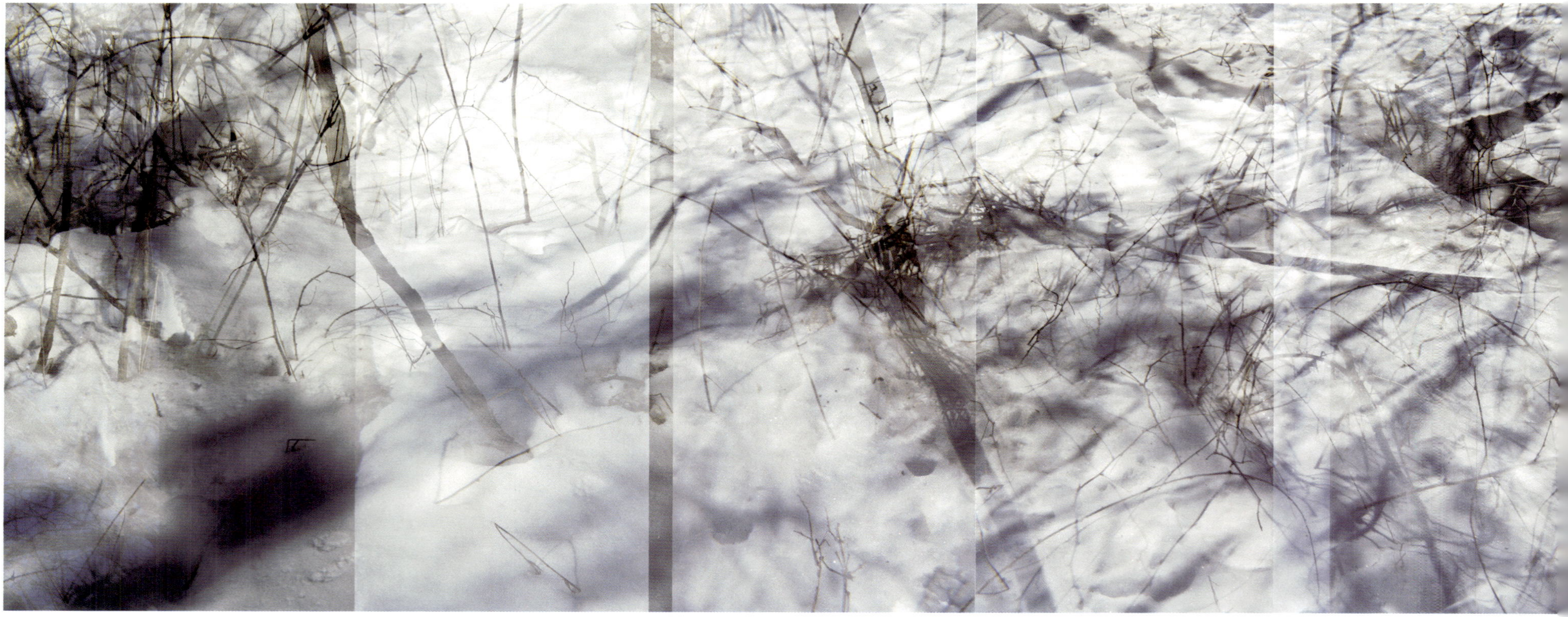

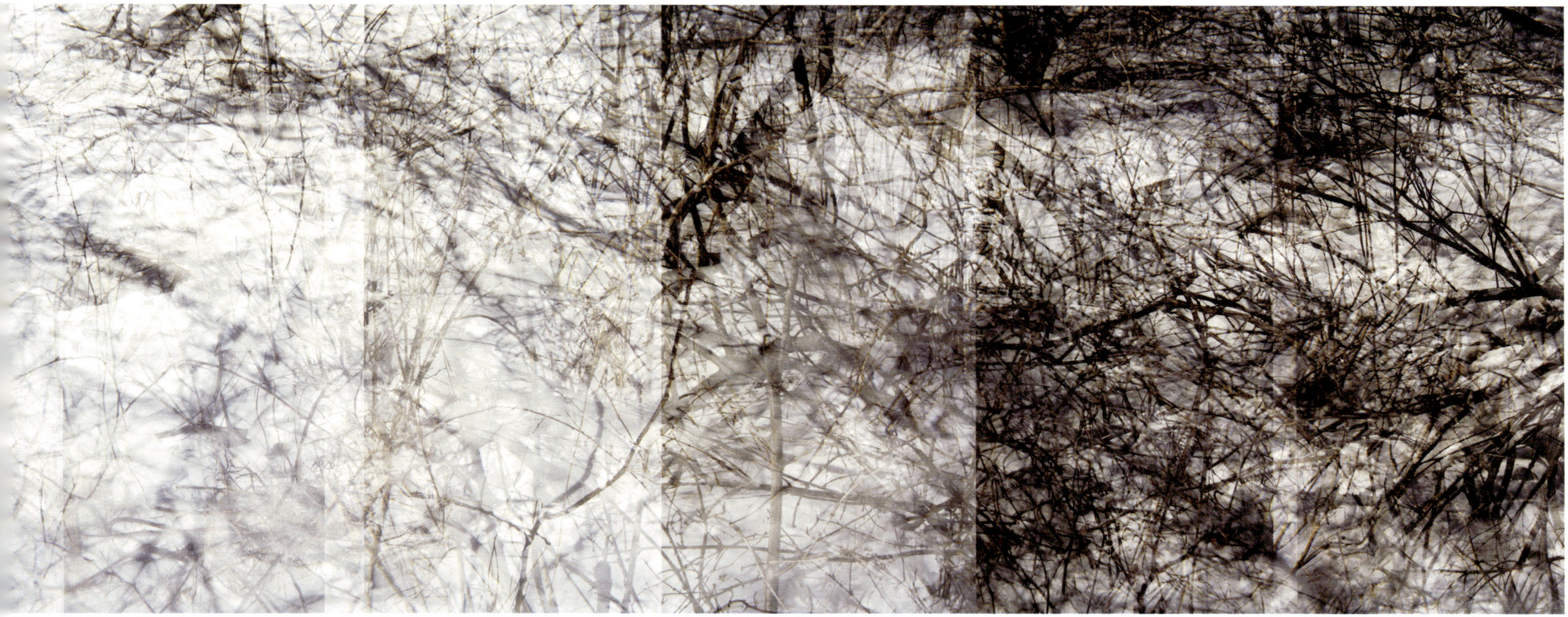

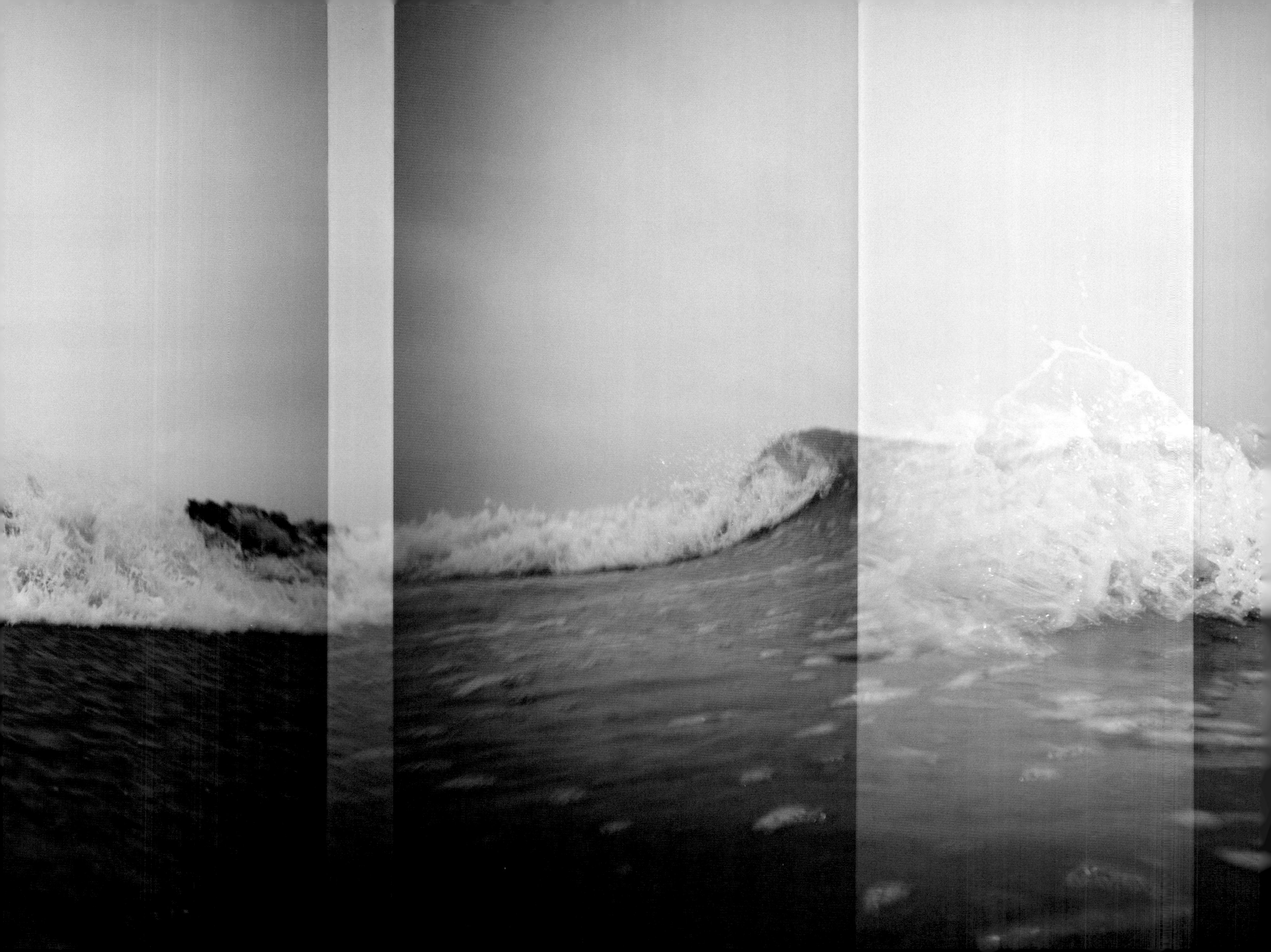

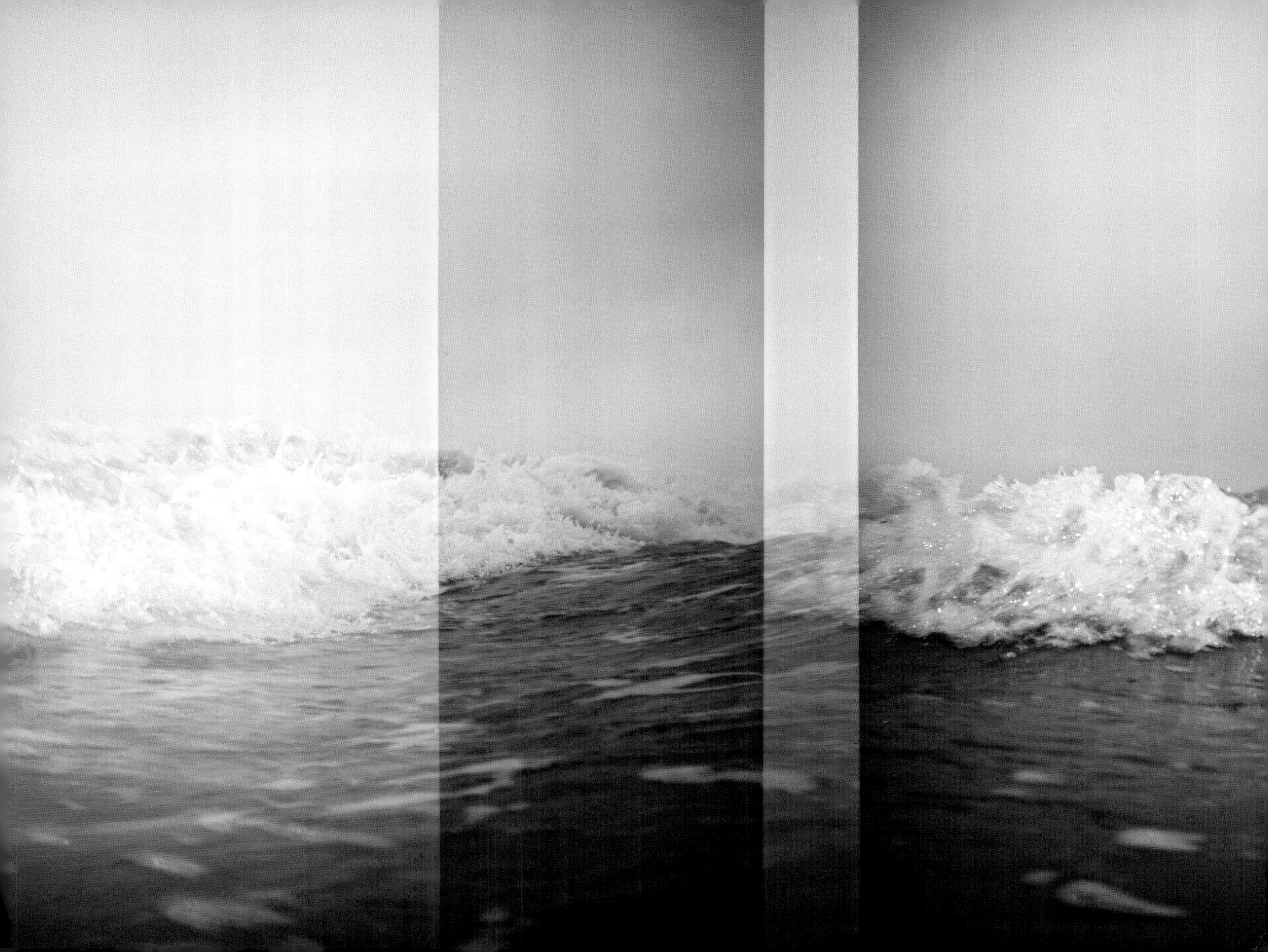

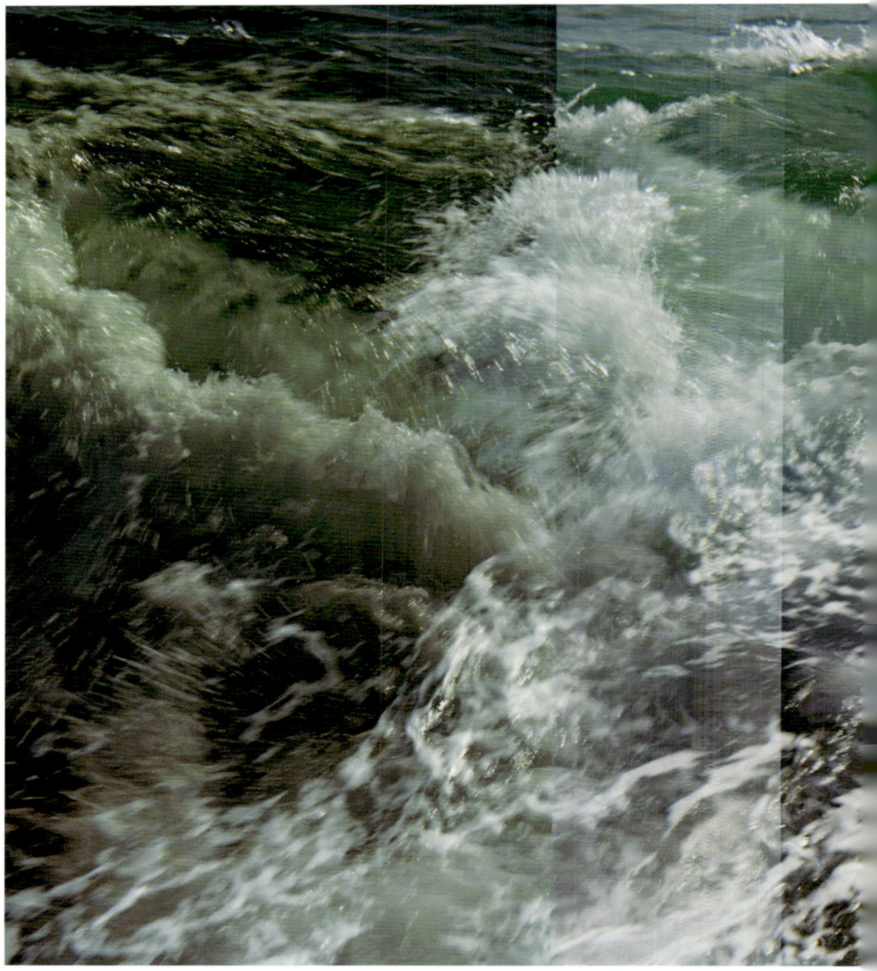

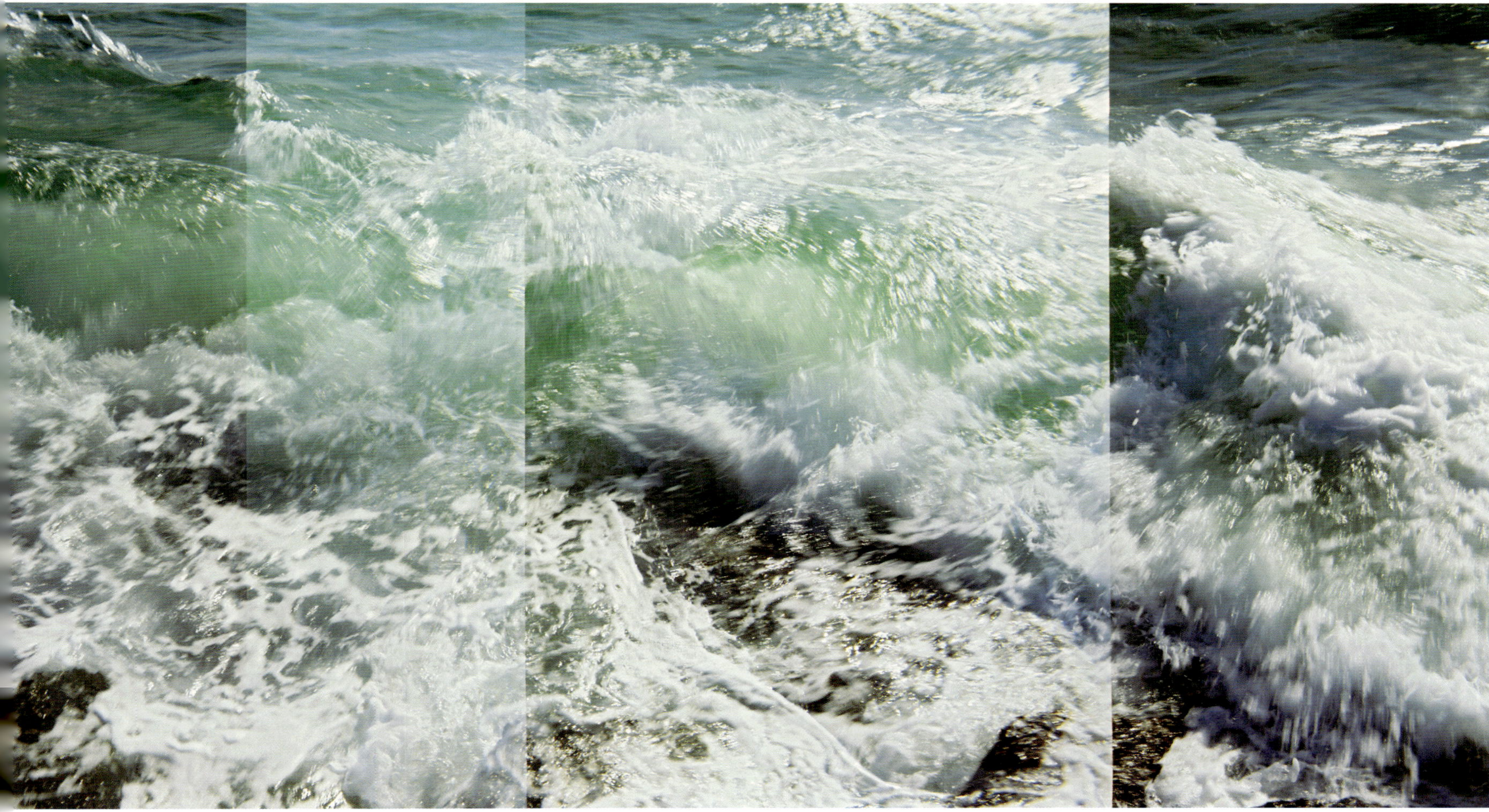

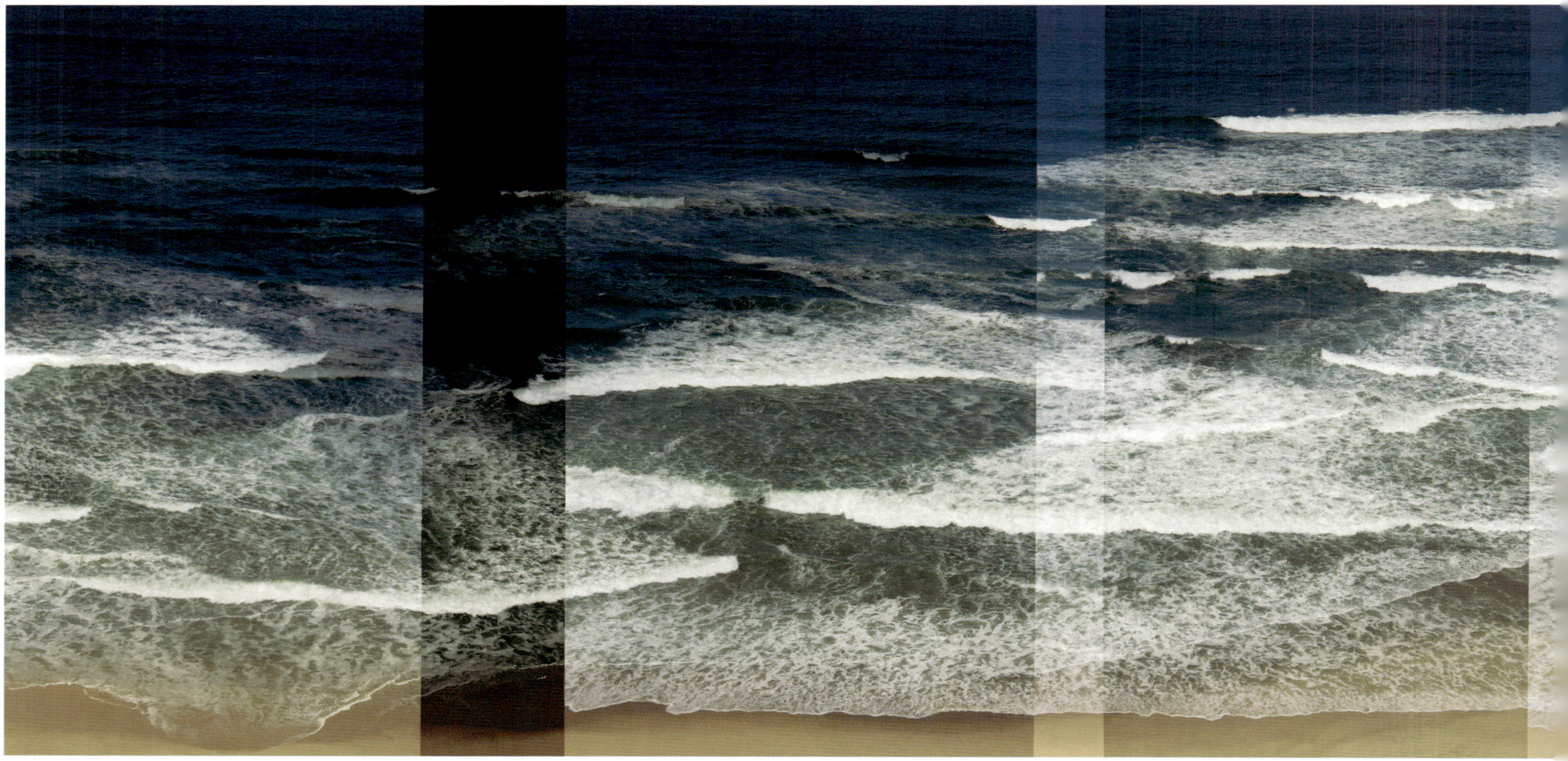

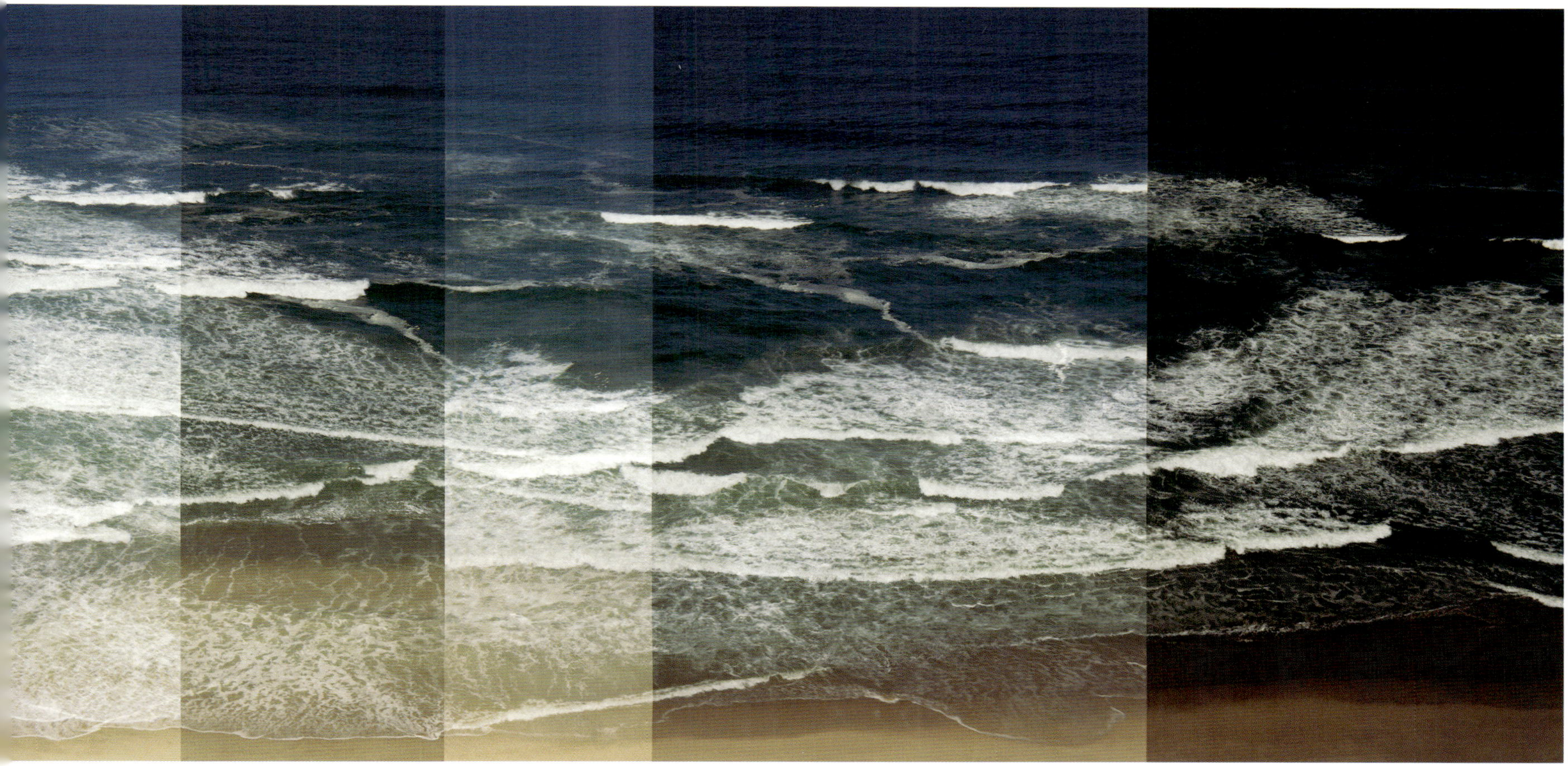

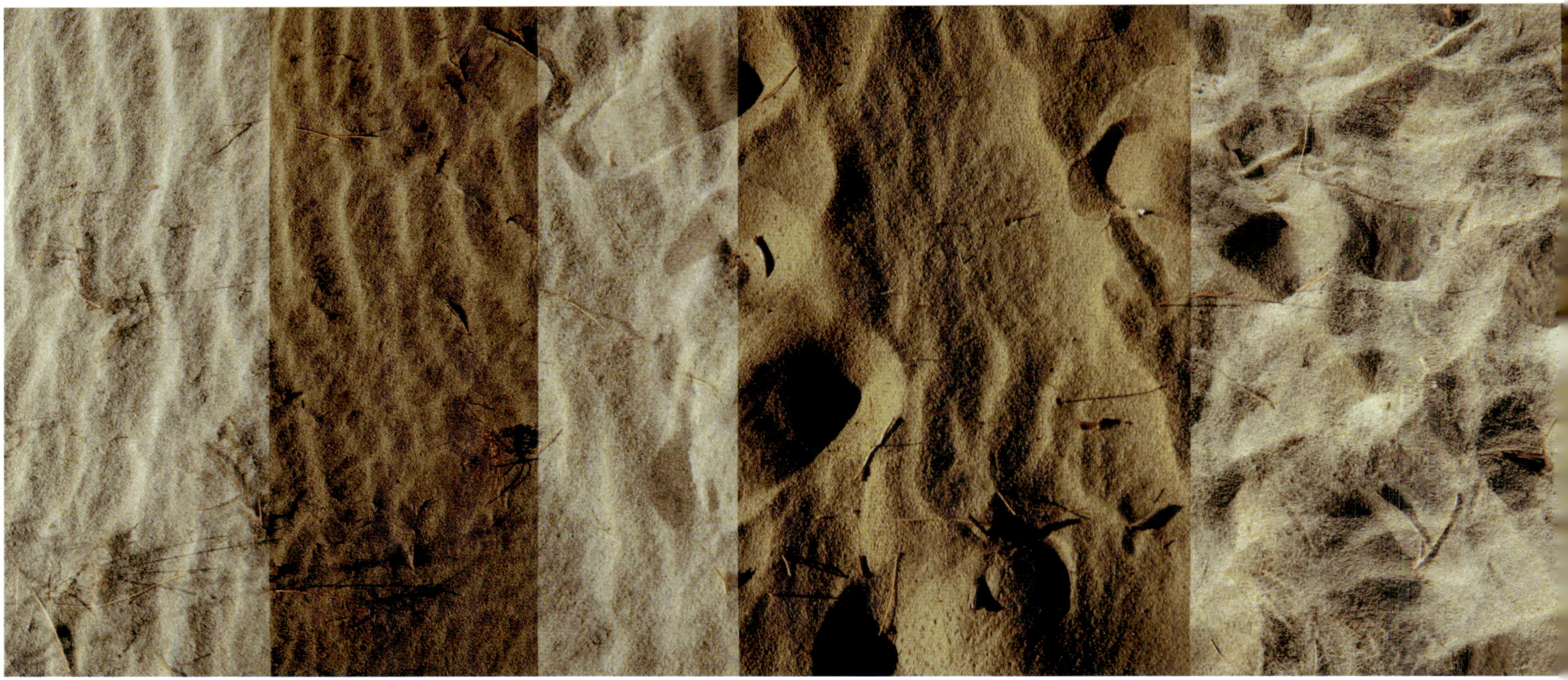

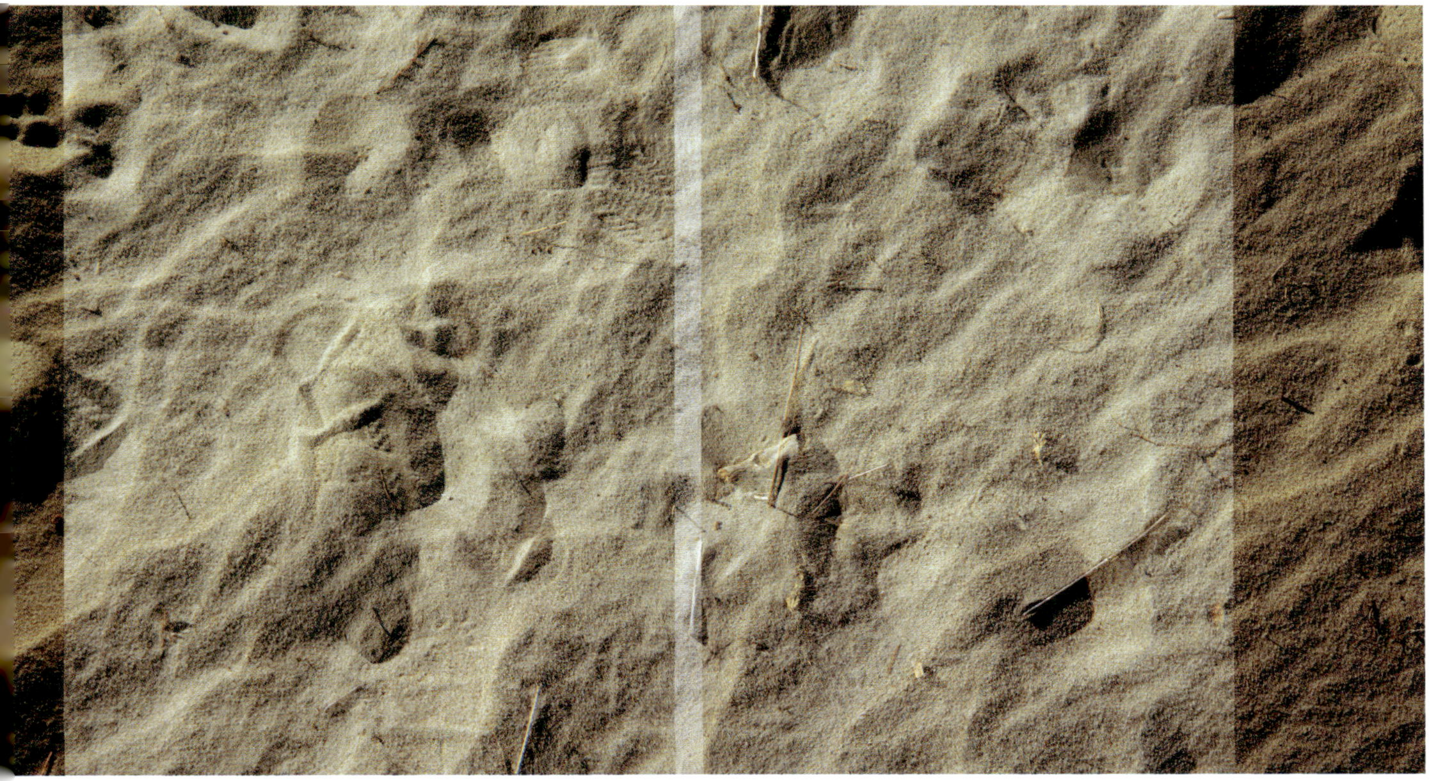

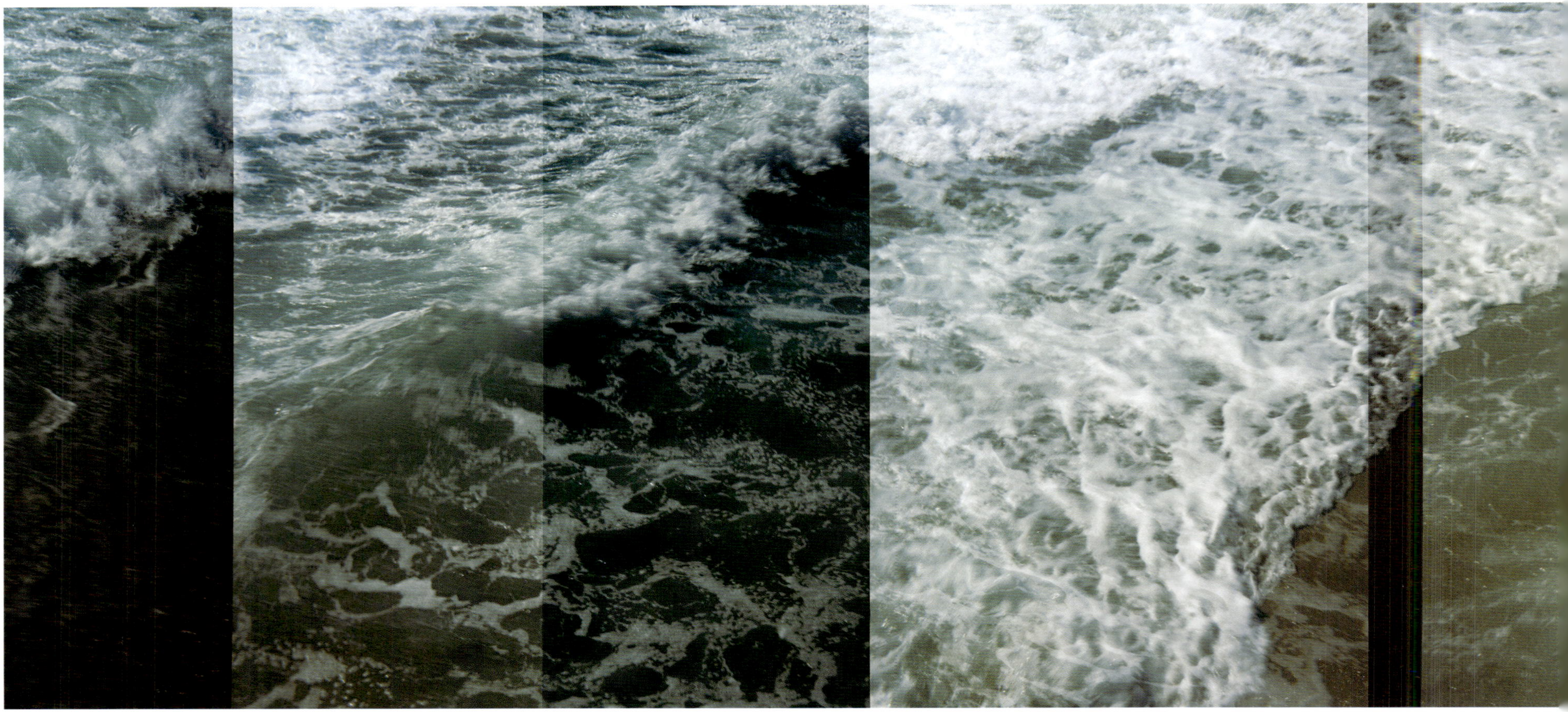

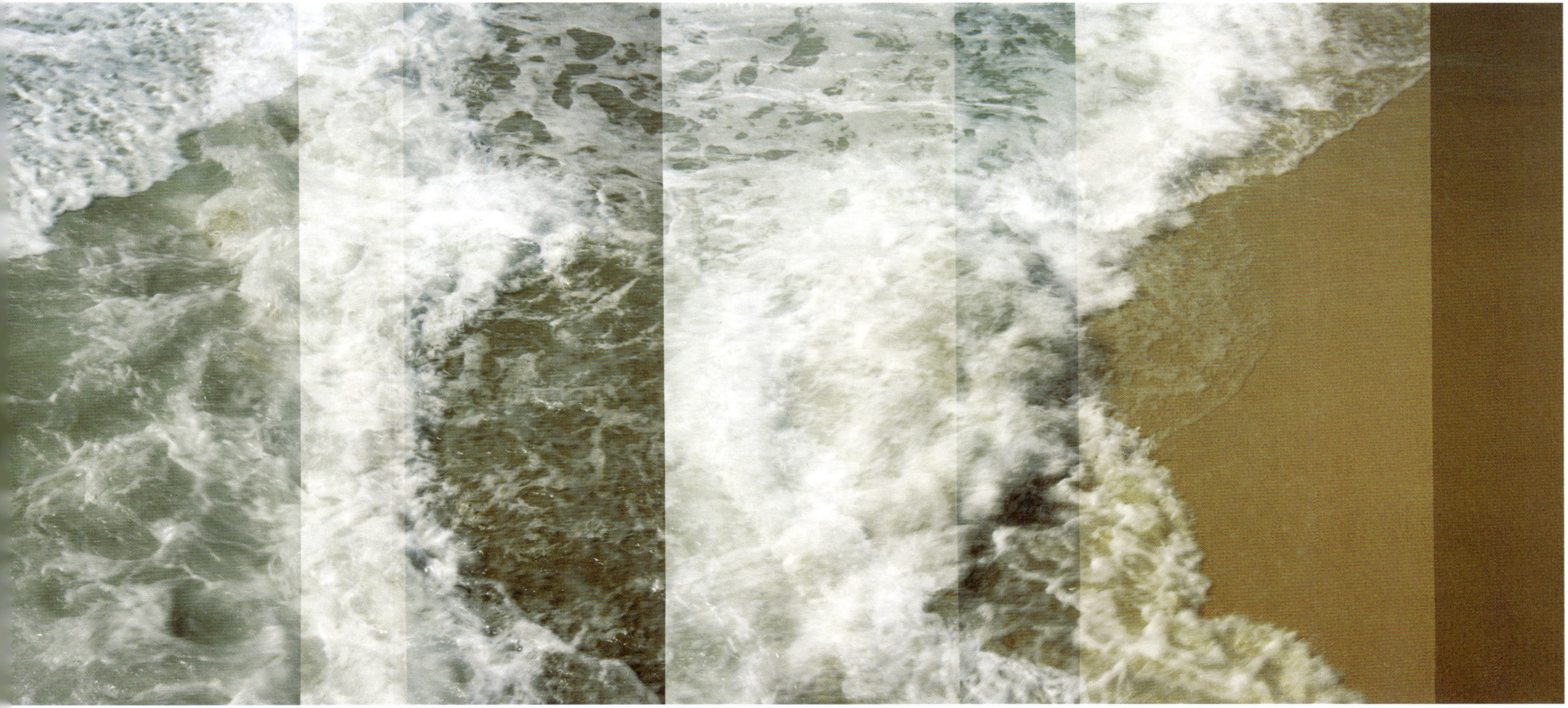

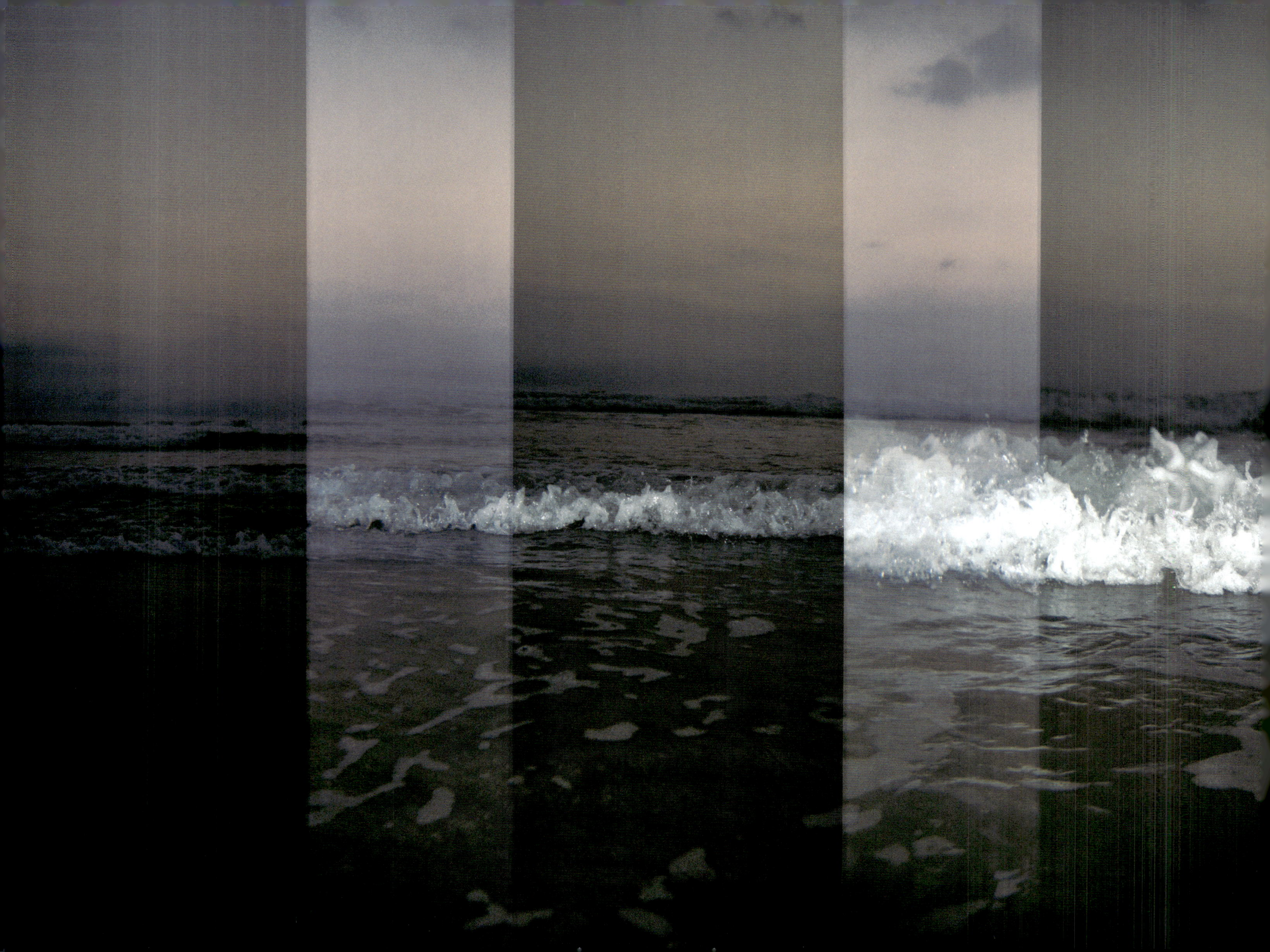

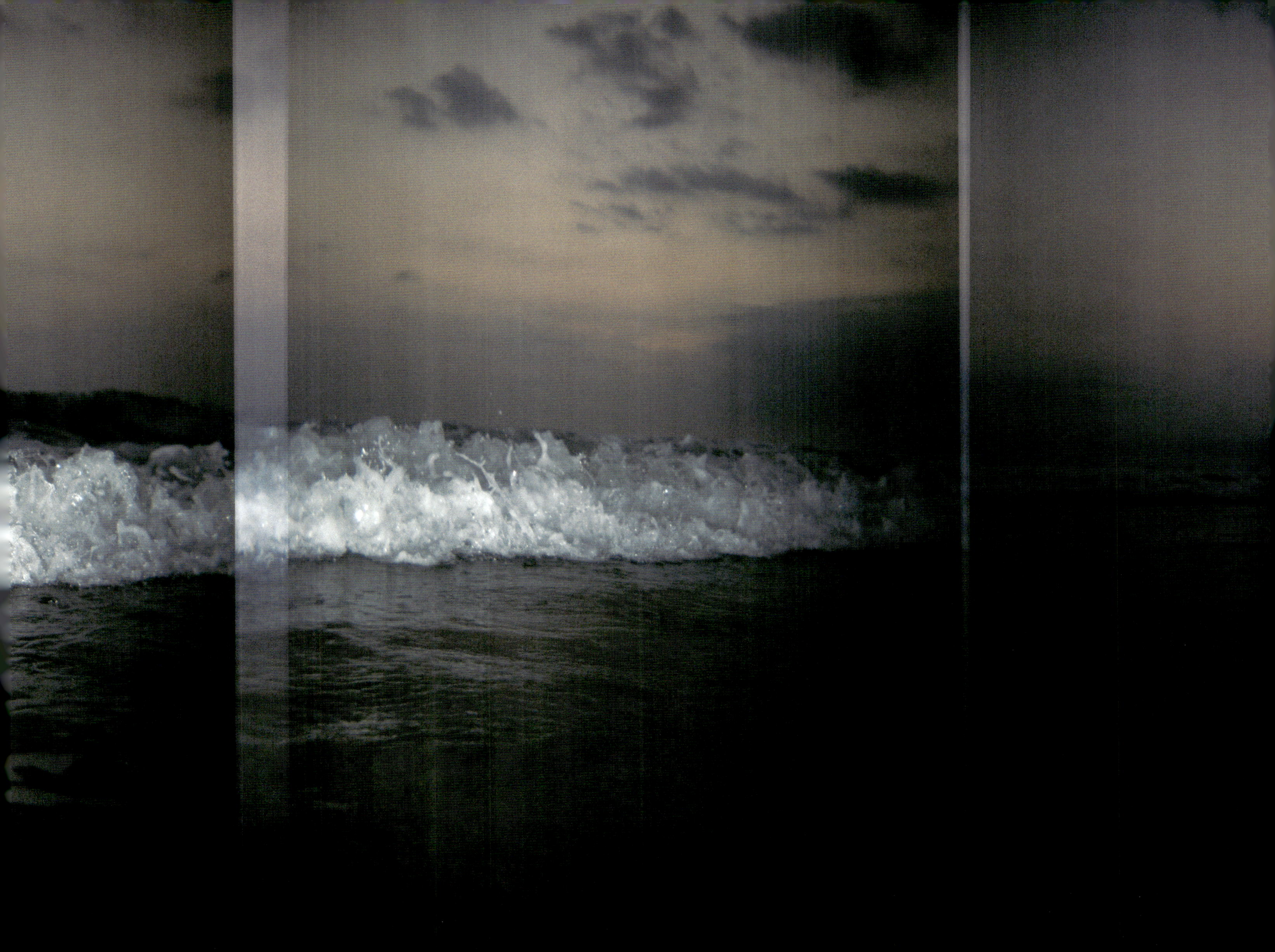

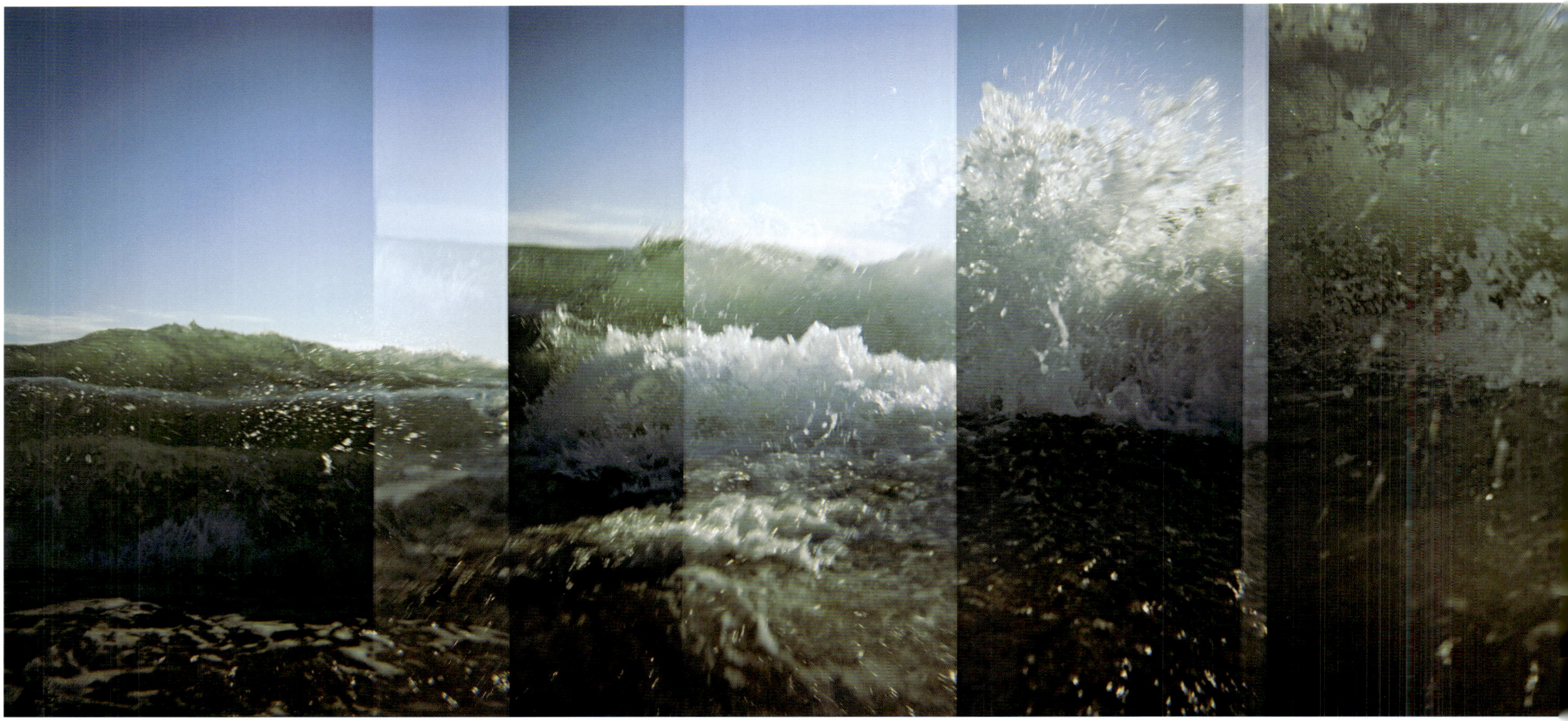

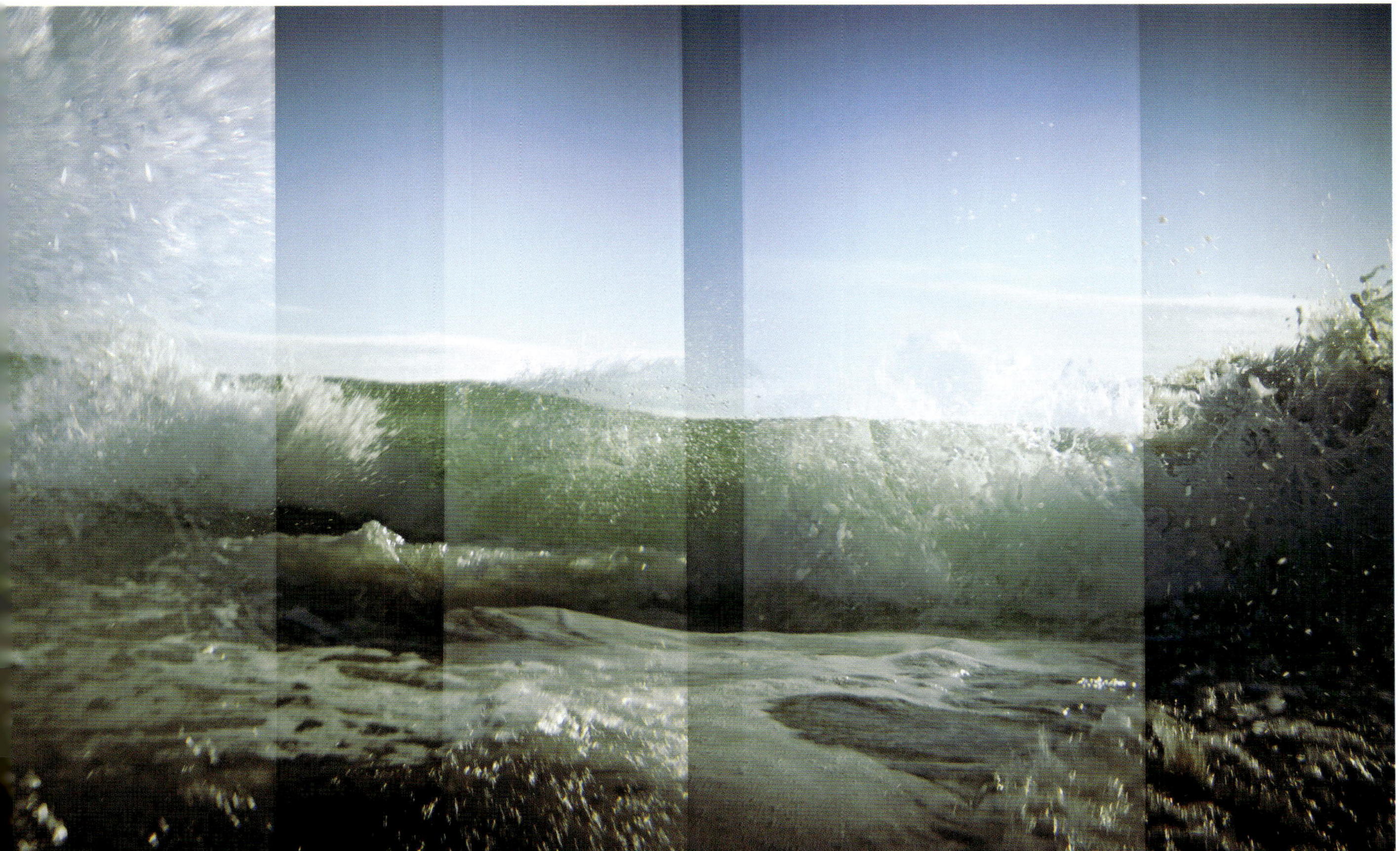

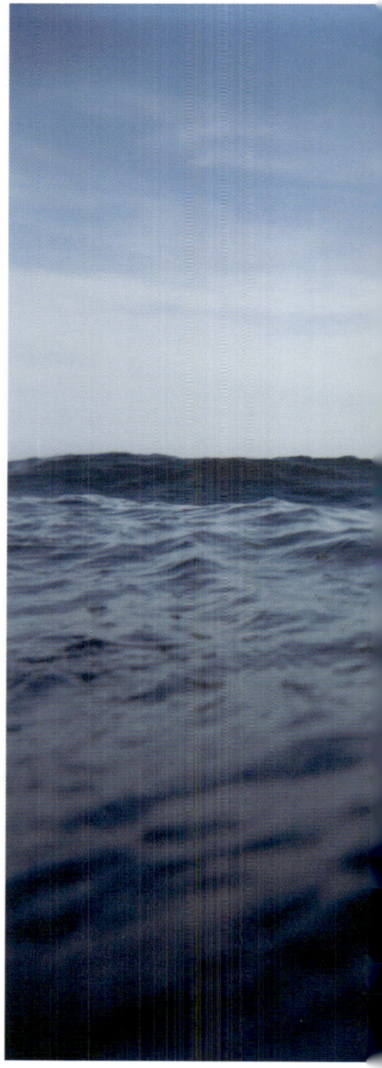

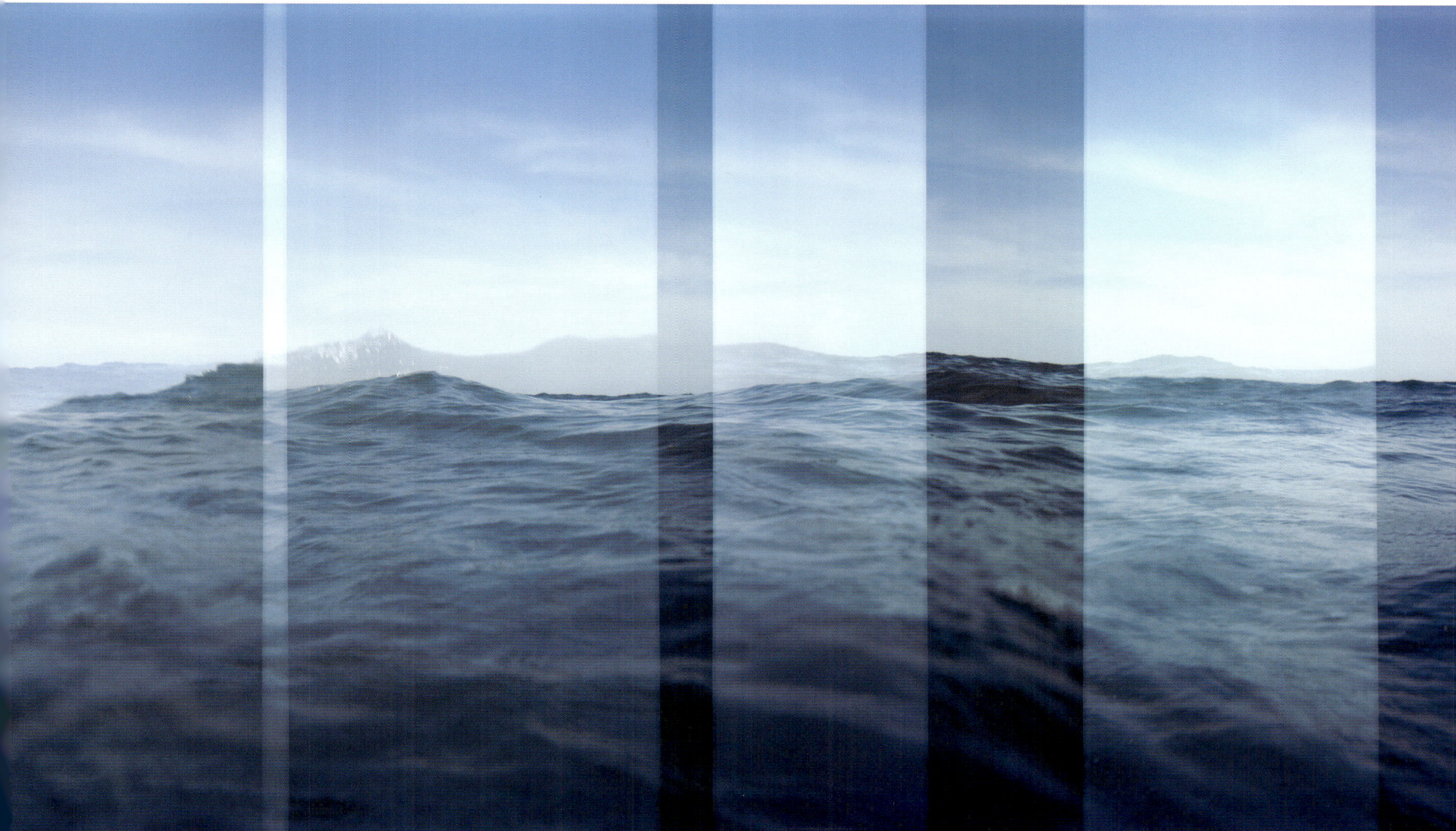

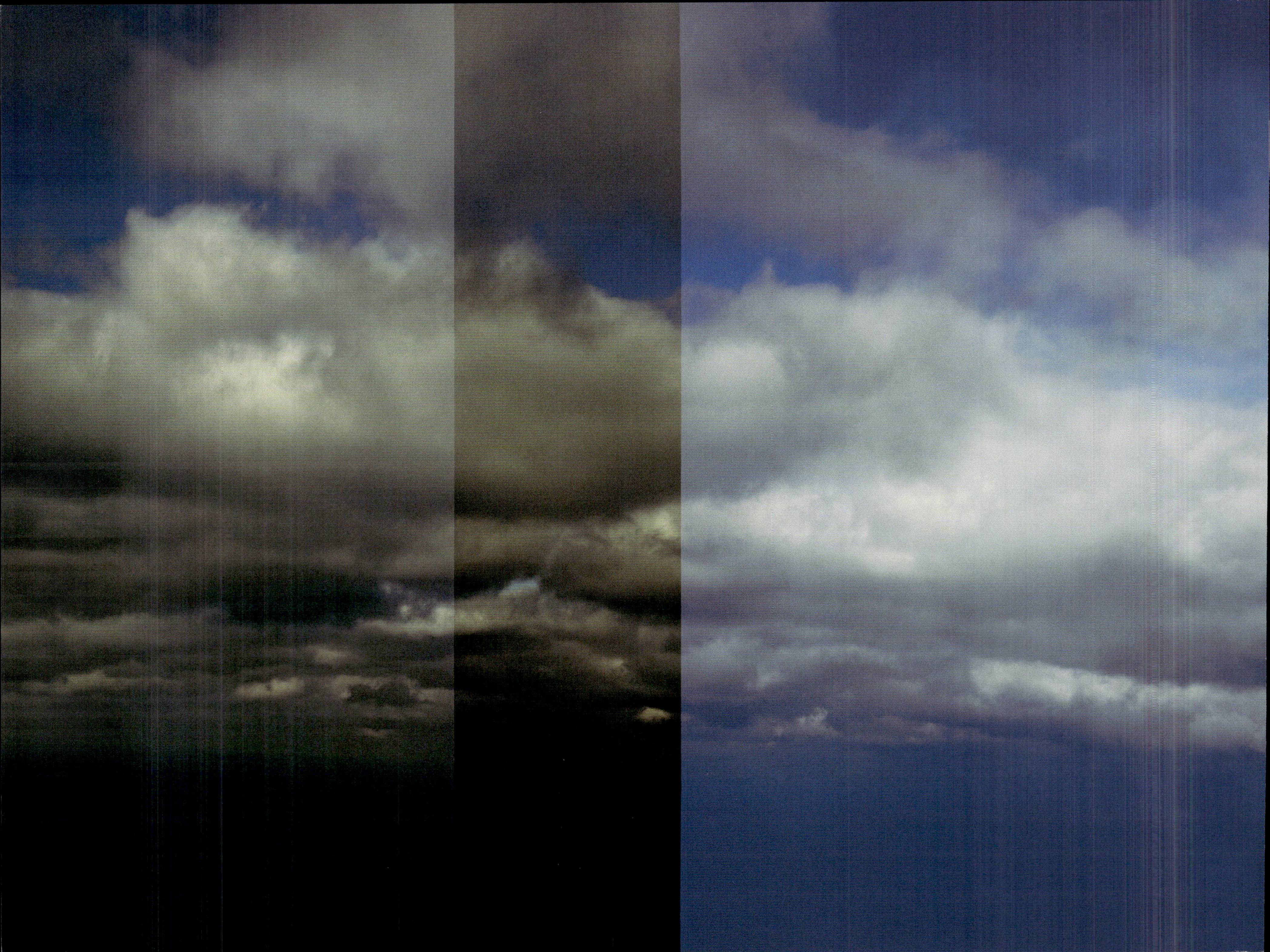

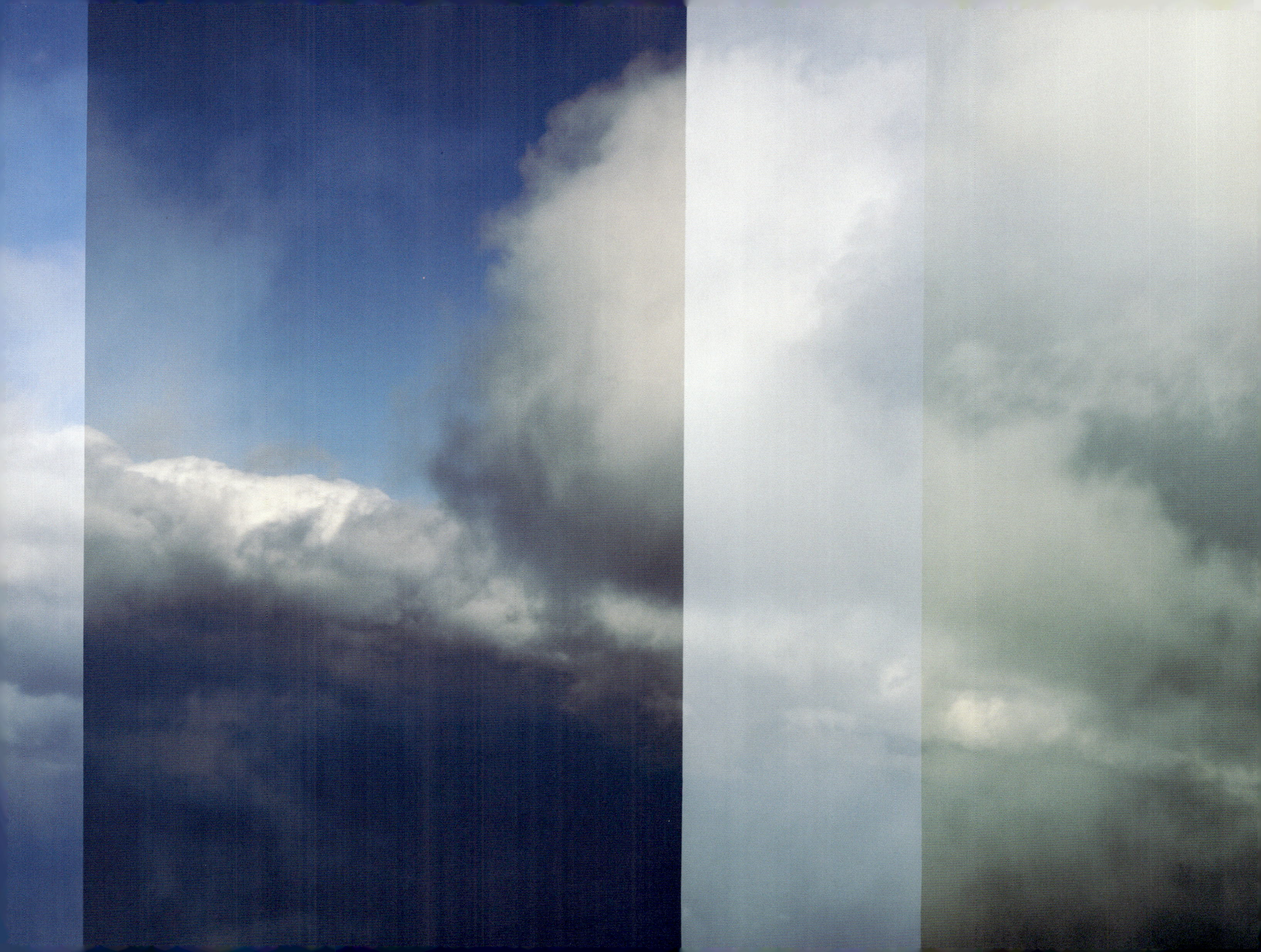

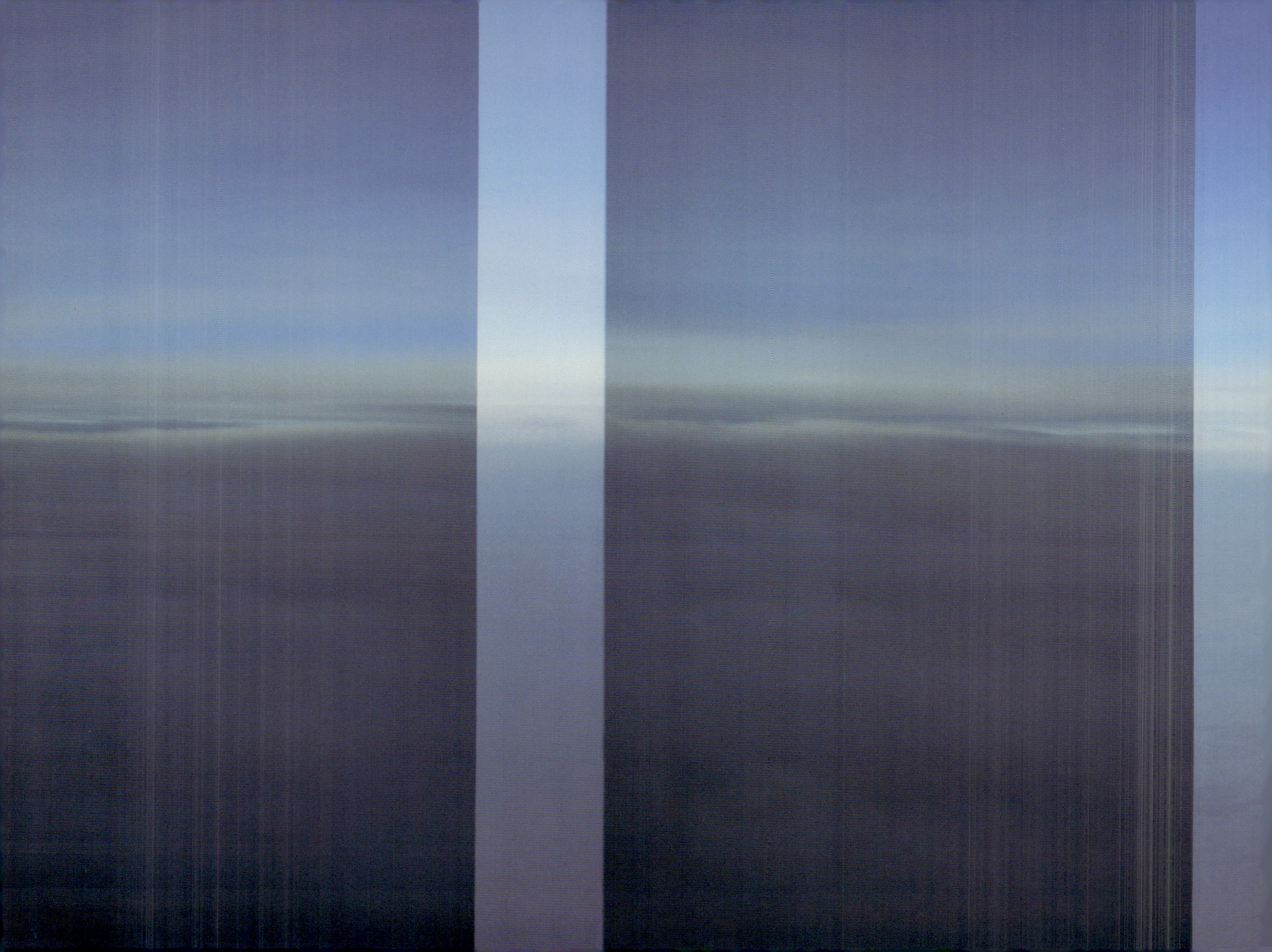

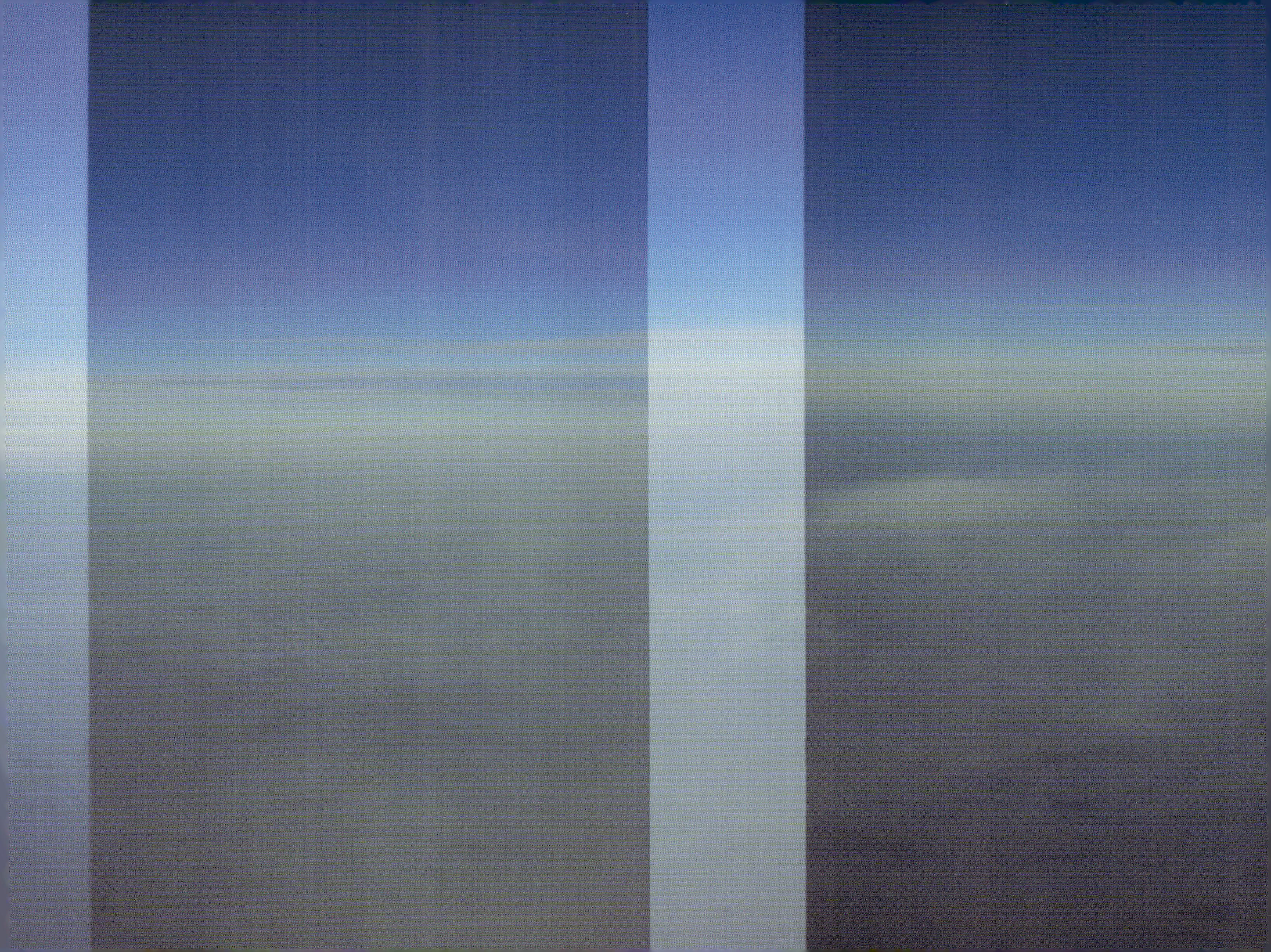

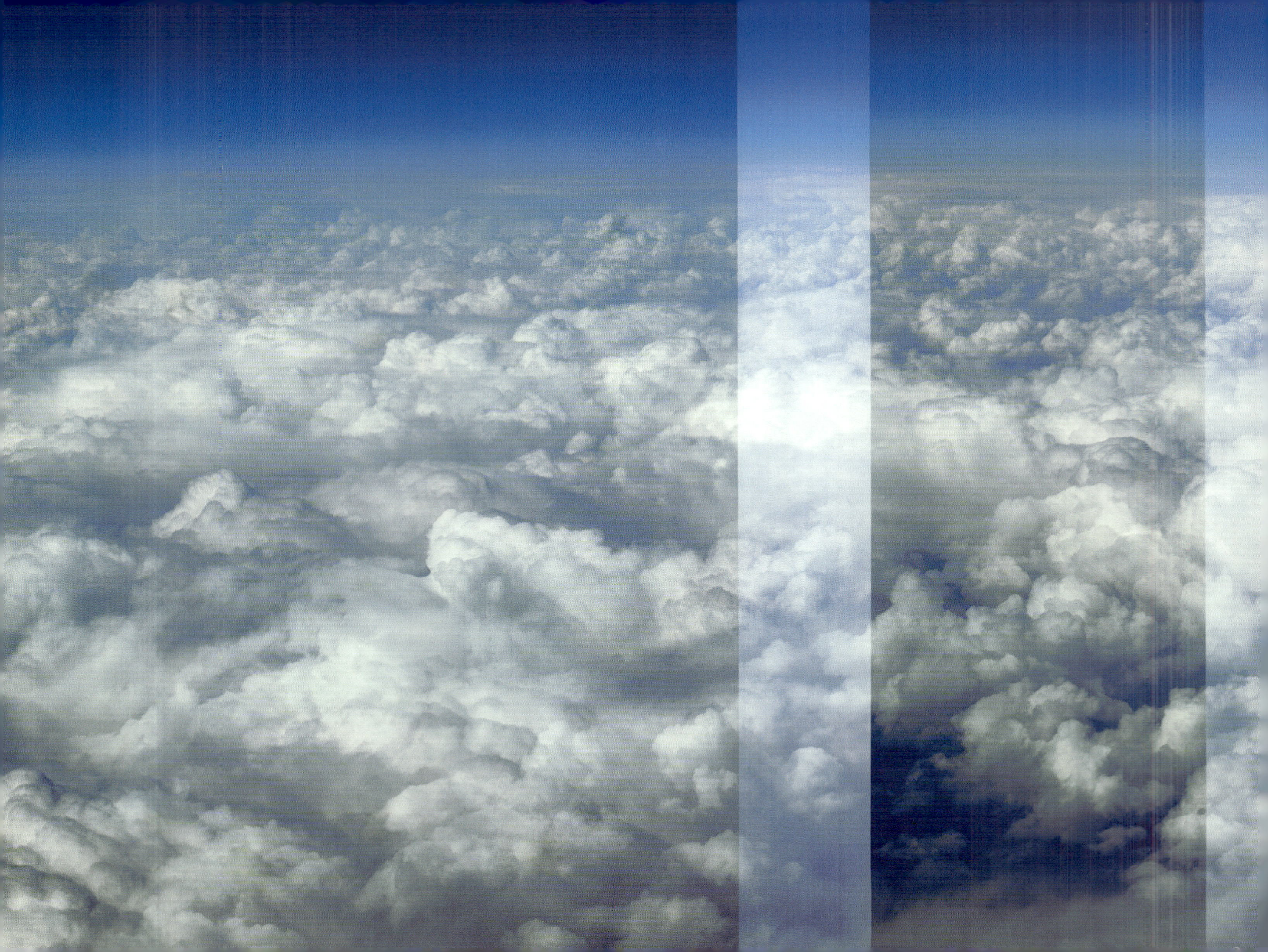

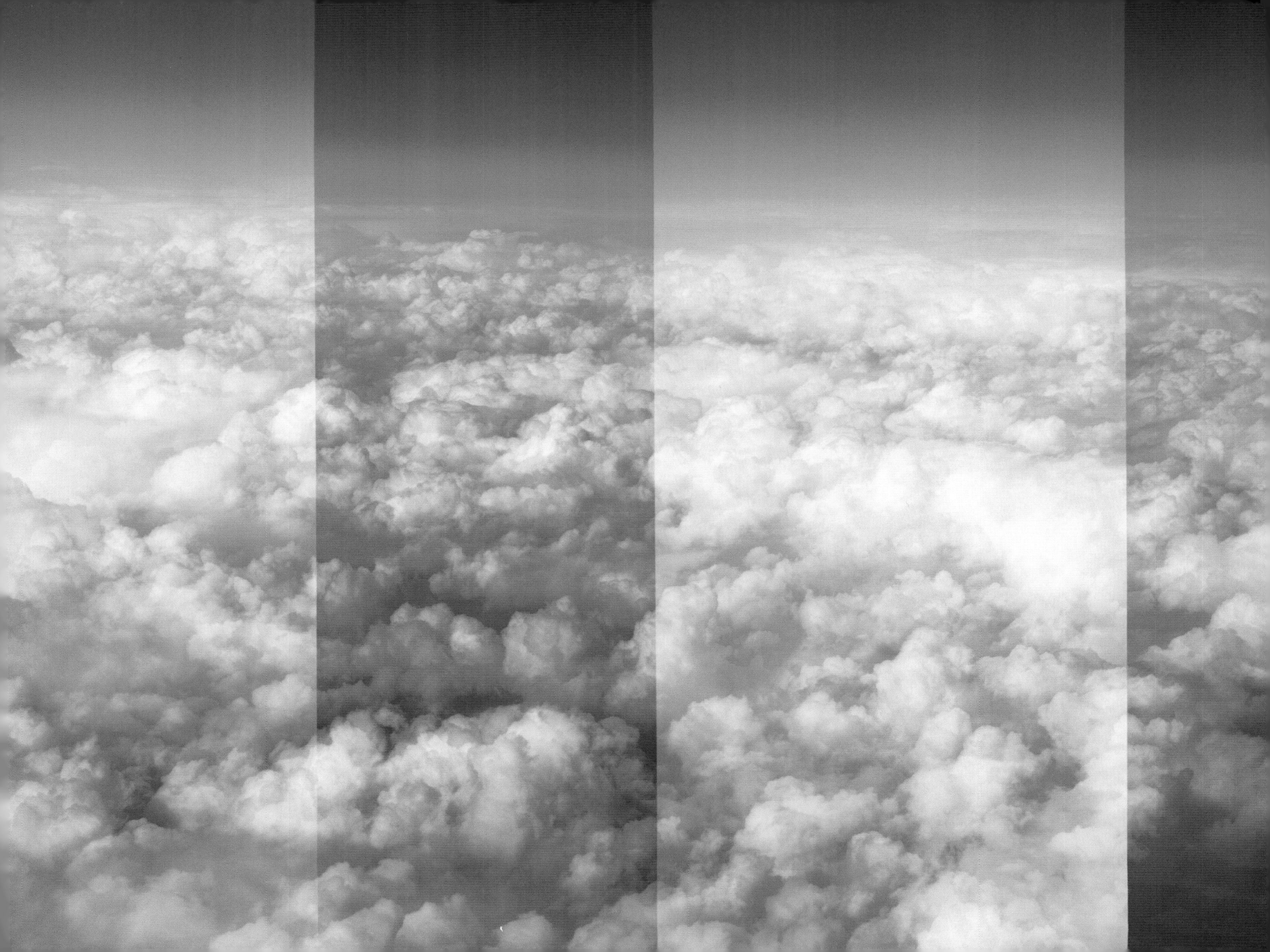

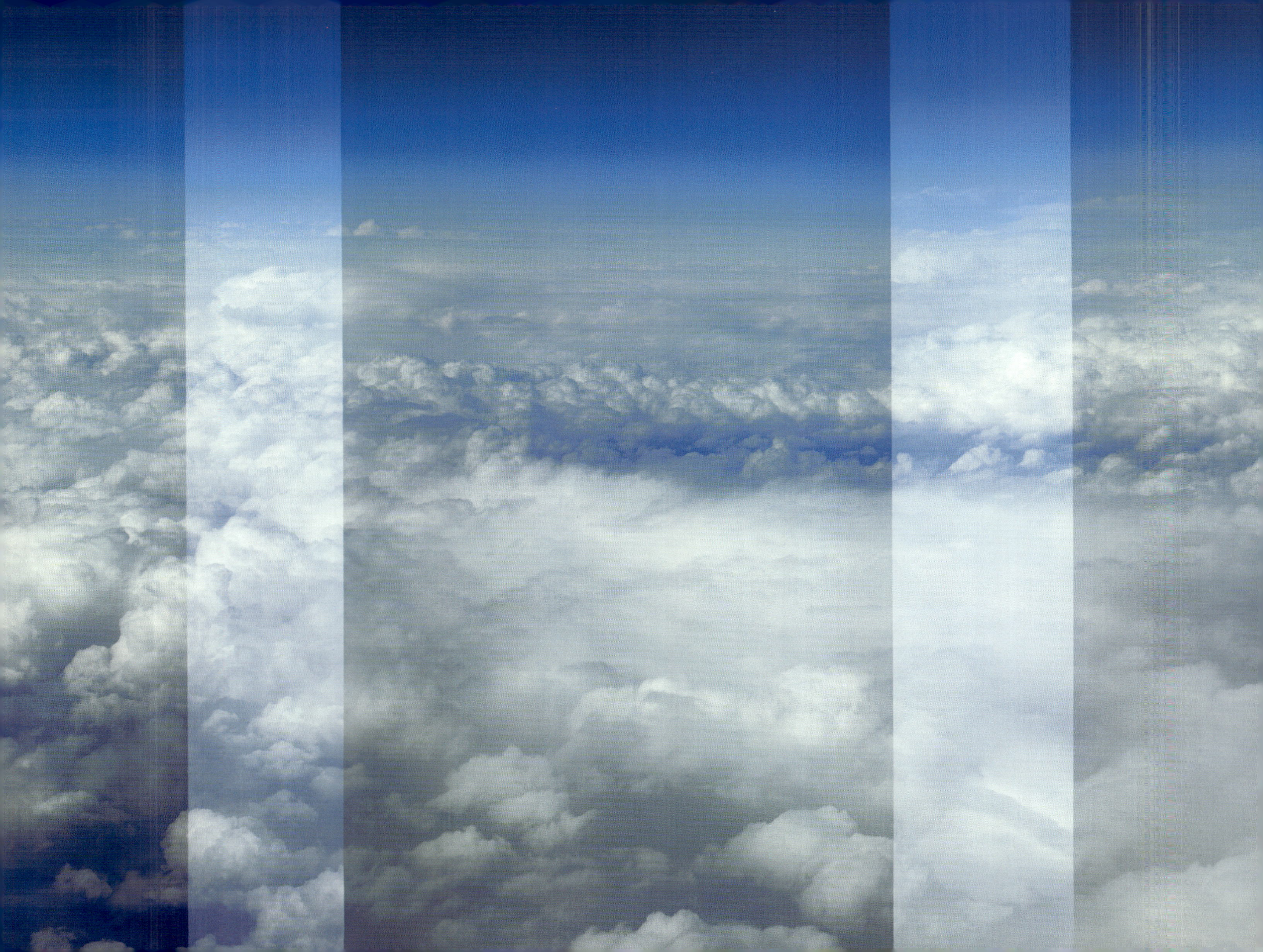

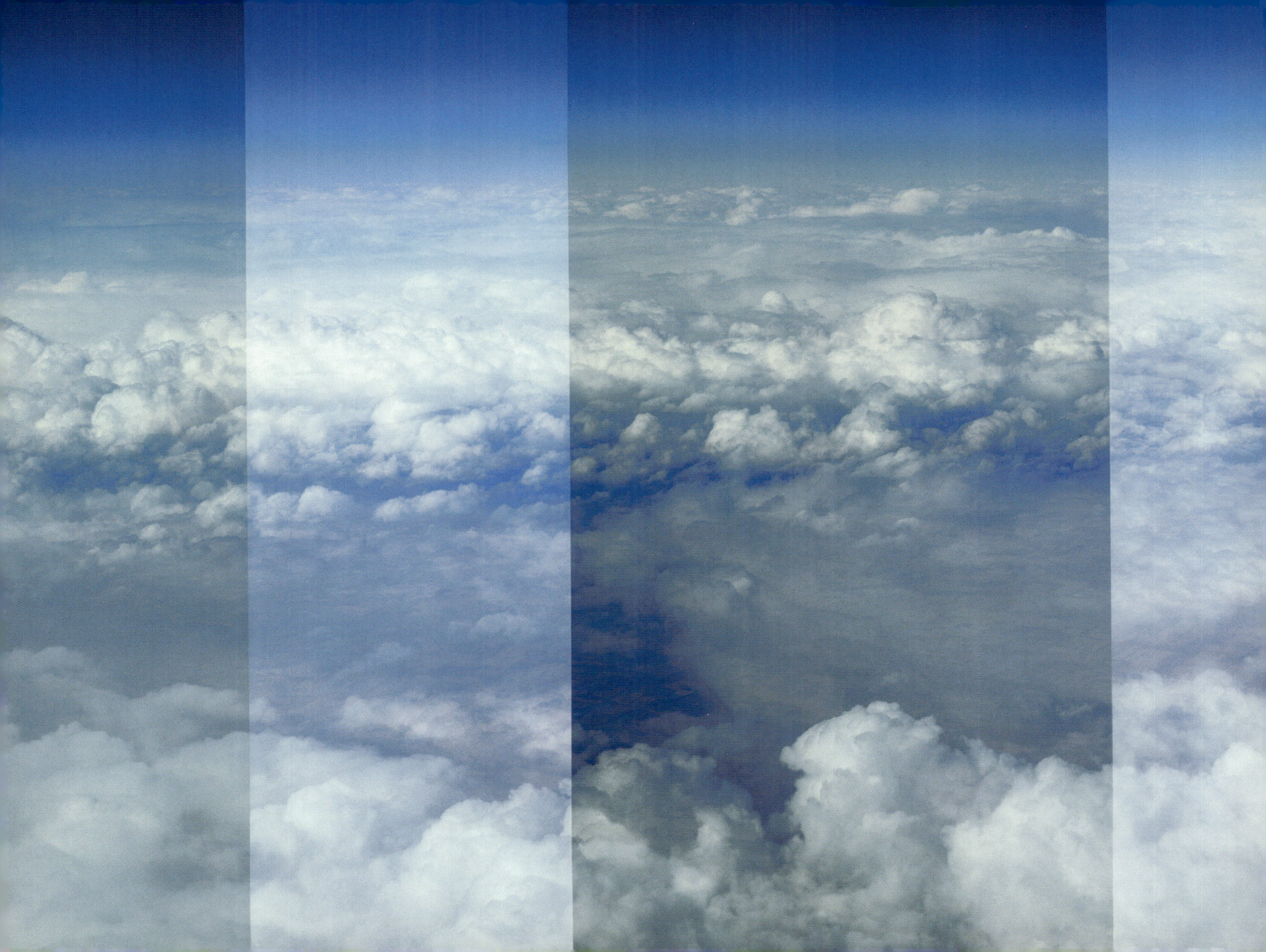

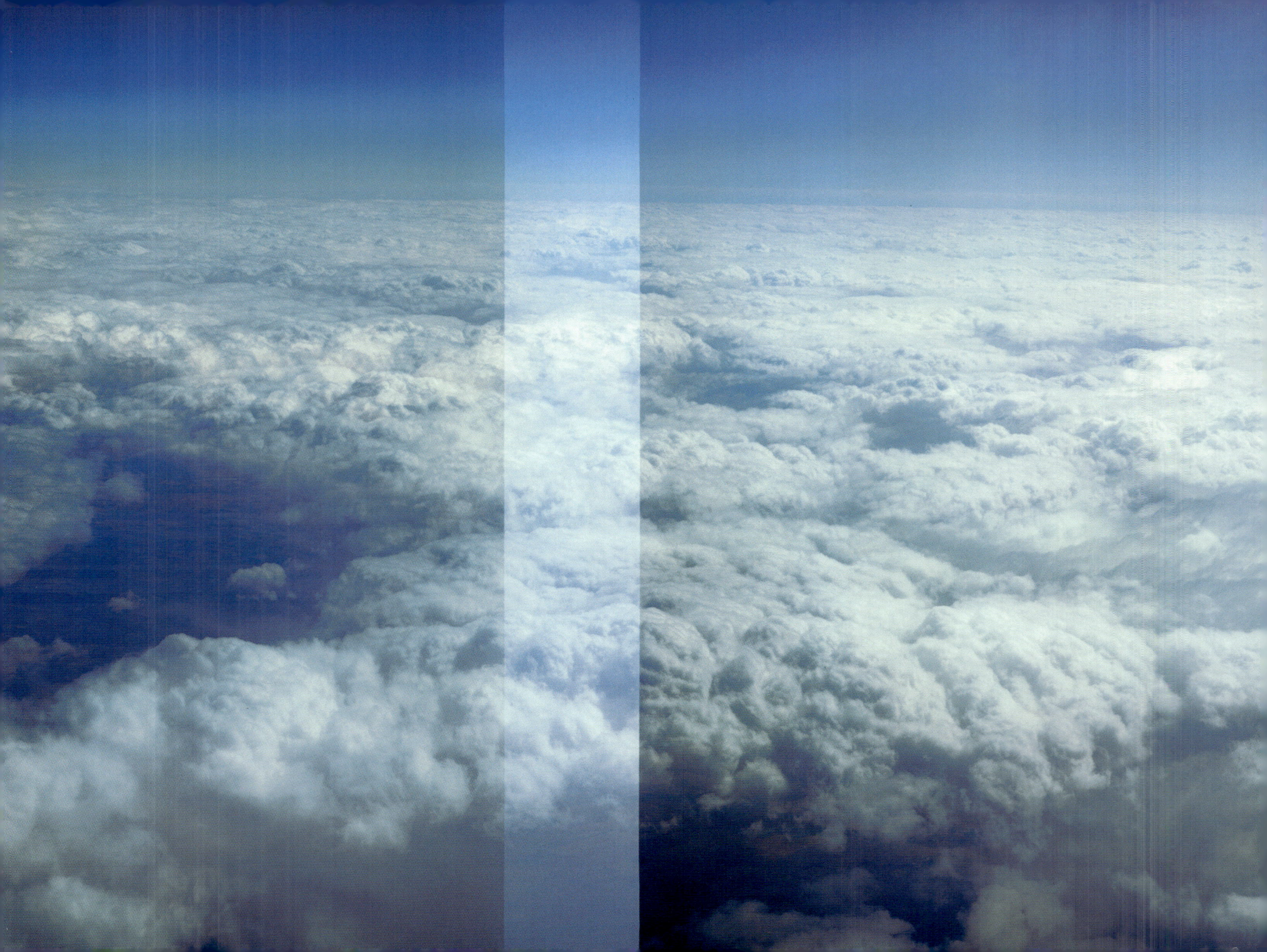

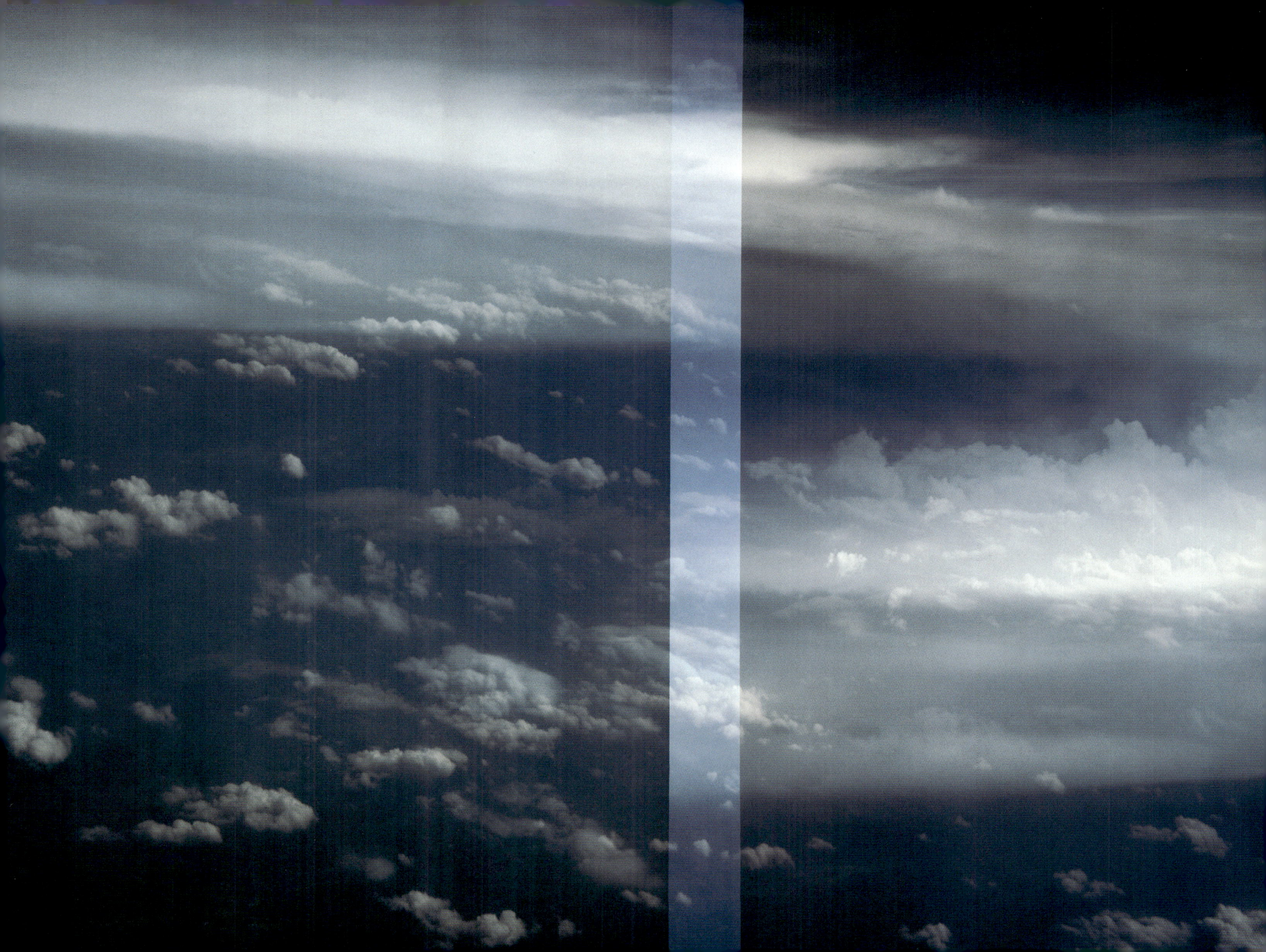

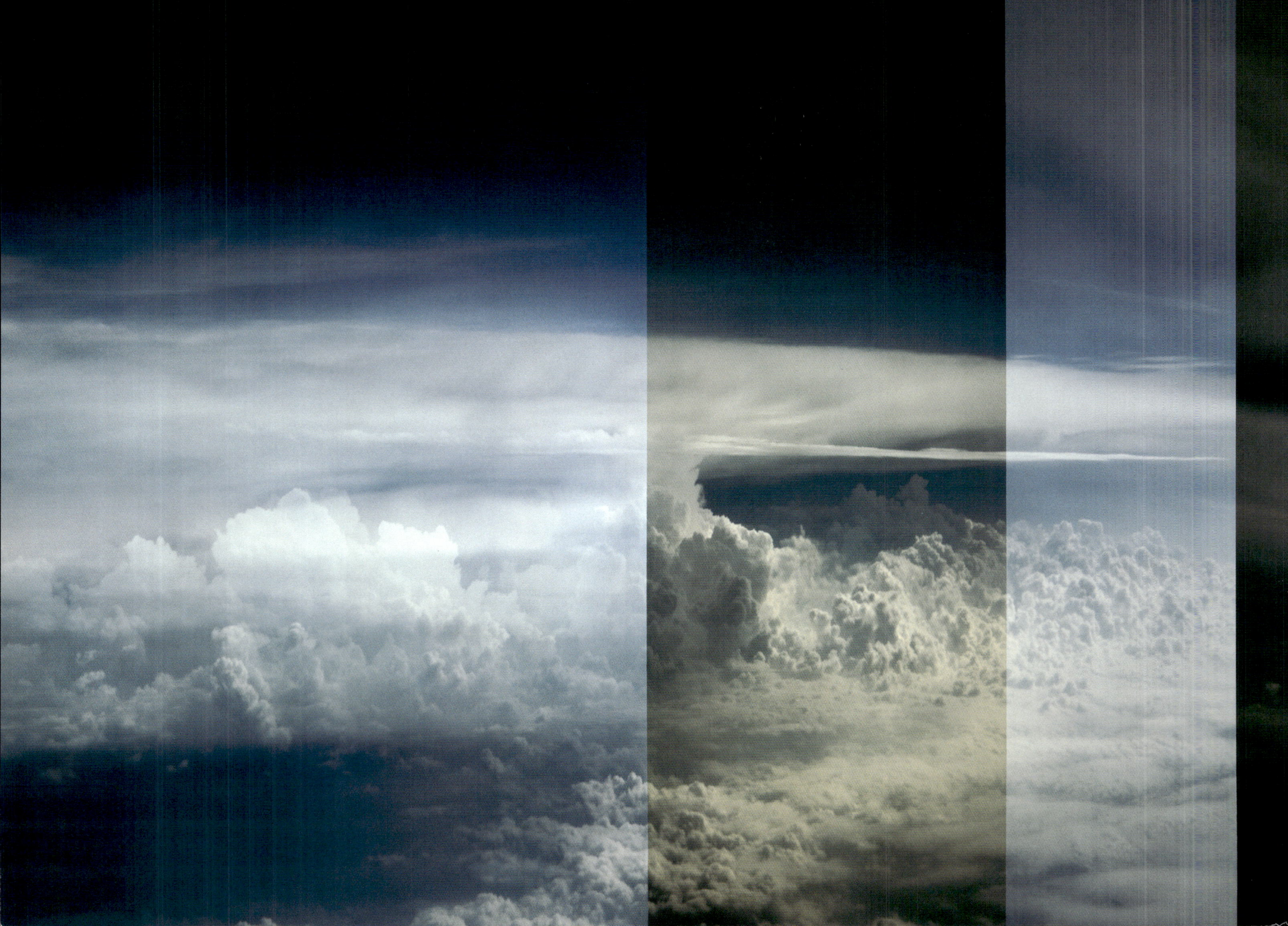

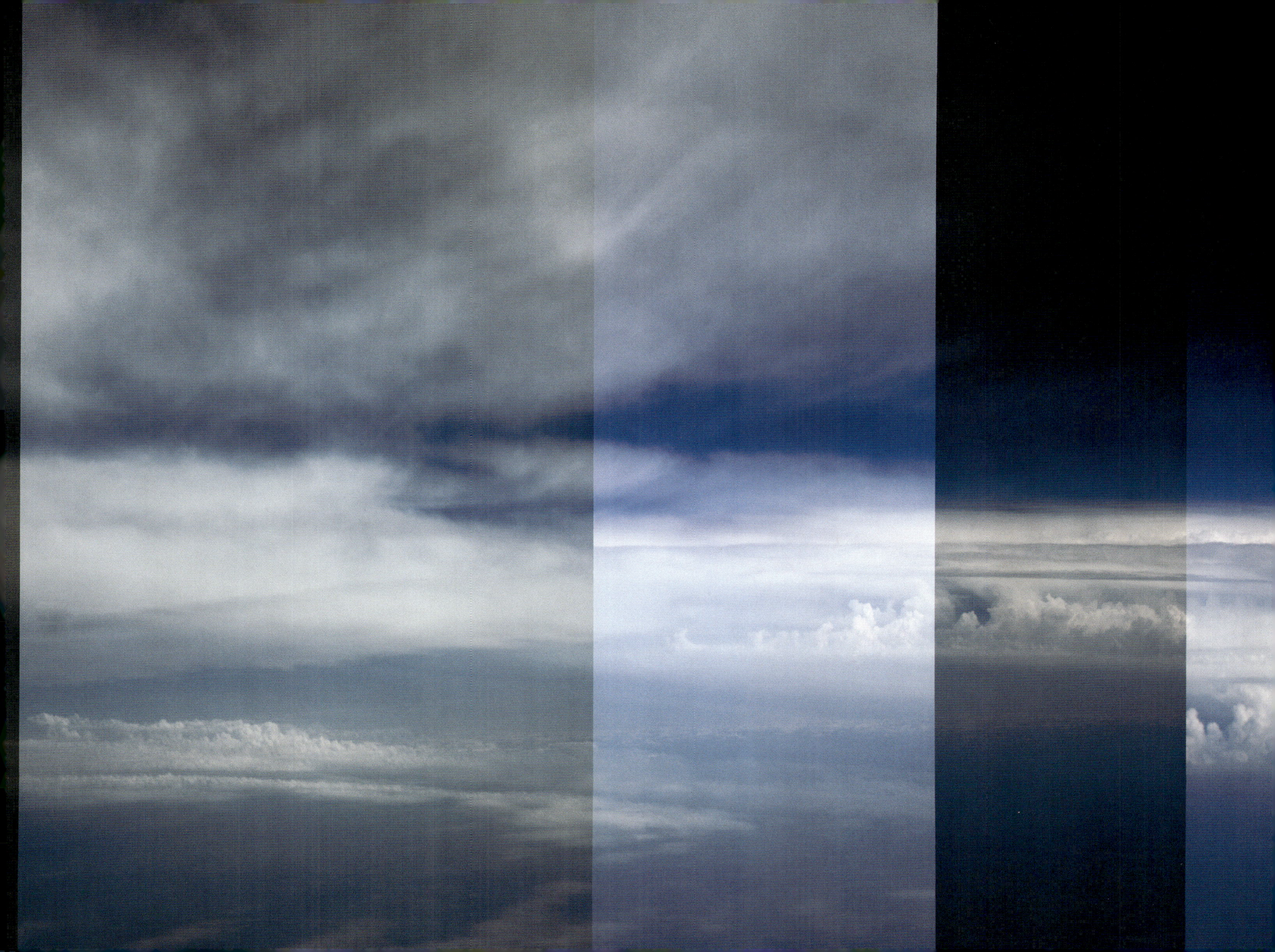

Eiren Caffall

Eiren Caffall was born in New York City, a year and a day after the first Earth Day, and raised in the Berkshire Hills of western Massachusetts.

Born to Urbanites with no memory of the country they'd adopted, she spent her childhood creating language for the landscapes she encountered. She layered the existing and the invented in solitary days spent reading and establishing imaginary countries in the forest.

As an adult, her work as a writer and musician is occupied with the examining the intersection of landscape, memory, and loss. Her songs and essays focus on layering the private territory of grief and joy with the public world of one's surroundings. They touch on the isolation of the urban consciousness in nature and the longing for the natural in urban life.

She is currently a freelance writer living and working in Chicago, and has written for travel guides, created book reviews for *Punk Planet Magazine*, taught creative writing workshops to students from ages 4-18, and administrated arts education programs for organizations like Lookingglass Theatre Company.

She has recorded several albums of original music; her latest was *Civil Twilight*, completed and self-released in 2004. She has performed solo and with backing bands in Chicago and on tour, opening for Vic Chesnutt, Simon Joyner, and Richard Buckner.

She is at work on a book-length collection of essays. Eiren lives on Chicago's north side with her husband Jason and son Dexter, where they are only two blocks from Lake Michigan.

Derrick Jensen

Derrick Jensen is the acclaimed author of thirteen books, including *A Language Older Than Words*, *The Culture of Make Believe*, and *Endgame*. Author, teacher, activist, small farmer, and leading voice of uncompromising dissent, he has been hailed as the philosopher poet of the environmental movement.

Writes *Publishers Weekly*, "Jensen paints on a huge canvas an emotionally compelling and devastating critique of the intellectual, psychological, emotional and social structure of Western culture."

His premise is as profound as it is persistent: industrial civilization is inherently unsustainable. It will always require violence to biotic and human communities. And it will create a culture where trauma is normalized, where living beings become objects, and where the only relationship left is one of domination.

Jensen's intricate weaving together of history, philosophy, environmentalism, economics, literature, and psychology has produced powerful arguments that demand attention in the tradition of such important writing as Herbert Marcuse's *Eros and Civilization* and Brigid Brophy's *Black Ship to Hell.* He vividly imagines an end to technologized, industrialized civilization and catapults this discussion into a passionate call for action. Jensen guides us toward concrete solutions by focusing on our most primal human desire: to live on a healthy earth overflowing with uncut forests, clean rivers, and thriving oceans that are not under the constant threat of being destroyed.

Jensen's writing has been described as "breaking and mending the reader's heart" (*Publishers Weekly*). He writes for *The New York Times Magazine*, *Audubon*, and *The Sun* magazine, among many others. He holds a degree in creative writing from Eastern Washington University and a degree in mineral engineering physics from the Colorado School of Mines, and has taught at Eastern Washington University and at California's Pelican Bay State Prison. He has packed university auditoriums, conferences, and bookstores across the nation, stirring the crowds with revolutionary spirit.

Bridgette R. McCullough Alexander

Bridgette R. McCullough Alexander has served on faculty of the School of the Art Institute of Chicago in the departments of Art History, Theory and Criticism and Visual and Critical Studies. A 19th century French art historian and theorist, McCullough Alexander focuses on the Second Empire under the reign of Napoleon III and the intersection between the production of material culture and trade of luxury objects; the development of department store and haute couture; and the social history of the artist's model milieu as it relates to class/race selection and how these concerns were explored through artists of the mid-19th century. Additionally, her work focuses on the contemporary art market and the economic implications of race and money.

McCullough Alexander has lectured widely and has written for *H-France Review*, *American Artist Watercolor*, and *Time Out Chicago*. Her major study, *The Laure of Manet: [Re]Historicizing Race in Manet's Olympia*, is in the documentation center of the Musee d'Orsay in Paris. She is writing a comprehensive historical study on black female artist models from the 18th through the 20th centuries. And she is undertaking a book project on African American art collectors entitled *Black Market: The Art Collections of Prominent African Americans*. This four-part book series will document the history, collecting philosophies, and concepts of African Americans over the past 50 years.

For more than fifteen years, McCullough Alexander has worked as an art consultant for museums and other art nonprofit organizations in both New York City and Chicago. She is the founder of The BRMC Group, Inc., a scholarly-based art consultant firm.

Tim Hartford

Tim is a principal in the visual communications consulting firm of Hartford Design–a graphic design and brand identity practice based in Chicago. The firm produces print and digital solutions for some of the world's leading companies.

Tim is a graphic designer with more than twenty years of experience designing, managing and producing communications including interactive media, web sites, corporate identities, marketing collateral, literature systems, signage and interactive media, serving such clients as Abbott Laboratories, American Dietetic Association, Bank of America, Jones Lang LaSalle, OfficeMax, and Genzyme Corporation.

He has received national and international recognition for his work. It has received honors in many design competitions and has been published in numerous books and periodicals.

His educational background includes a Bachelor of Science degree in graphic design at Western Michigan University, where he was a member of the Honors College, as well as graduate-level studies in communications.

Tim formerly served on the Executive Board of the American Institute of Graphic Arts, Chicago Chapter, and on the Board of Directors of Roadworks Productions Theater Company.

Doug Fogelson

Doug Fogelson uses an iconoclastic multiple exposure technique in order to depict our collective surroundings, producing imagery that reflects our own alien experience of nature, as well as the distanced perspective of the viewer. His images are the result of overlapping multiple exposures along the film within the camera at the time of shooting. Fogelson's art depicts a measured study of time, complexity, and space to be found when scenes correspond and multiply. The viewer becomes enmeshed in the variegated forms captured by the photographer's eye and via the machine of the camera.

Fogelson's process takes a stance between motion picture and still photography as his images are often shot with changing vantage points of the scenes depicted and overlapped along the film. Movement is displayed–both of photographer and subject–yet a cognitive sense of the subject in consideration is rooted in time and perhaps personal reflection or memory.

Born in Chicago in 1970, Doug Fogelson studied at Columbia College and the Art Institute of Chicago, receiving his BFA in 1994. He has been recognized by numerous publications including *Art News*, *Focus*, and *Photo District News*, and his work is included in prominent museum, corporate, and private collections. Fogelson exhibits in galleries, museums, and public spaces around the world.

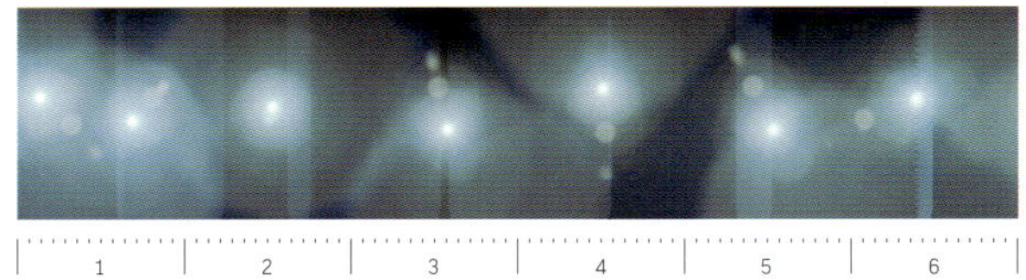

Sun Ra
18 x 85 inches,
Edition of 10,
Lambda C-print, 2007

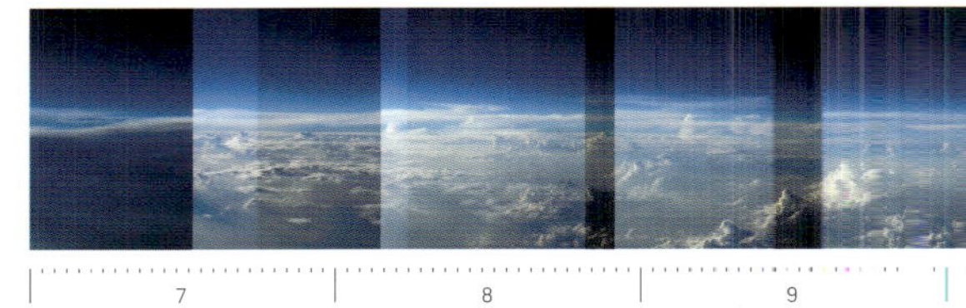

Returning
20 x 80 inches,
Edition of 10,
Lambda C-print, 2008

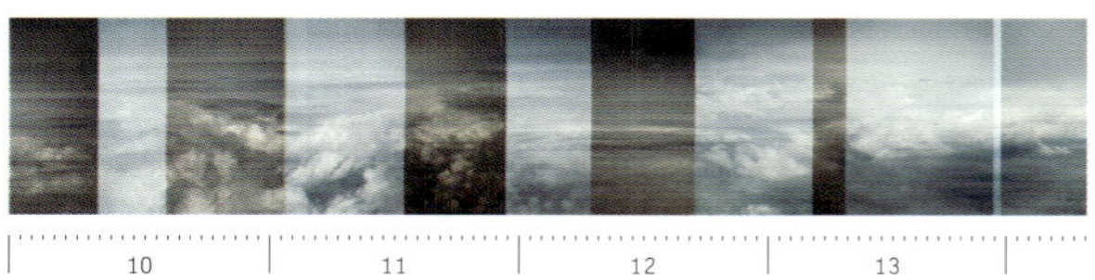

Passenger
18 x 96 inches,
Edition of 10,
Lambda C-print, 2006

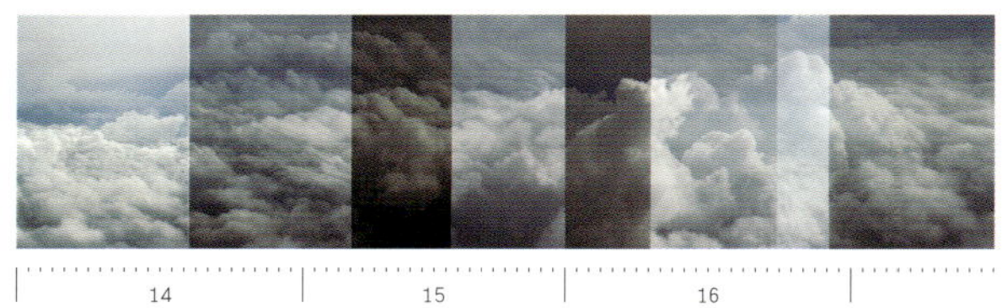

Dove Tail
20 x 83 inches,
Edition of 10,
Lambda C-print, 2006

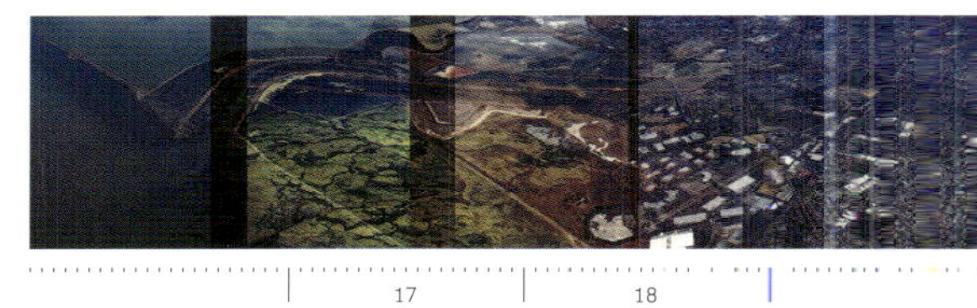

Bay (San Francisco)
20 x 84 inches,
Edition of 10,
Lambda C-print, 2006

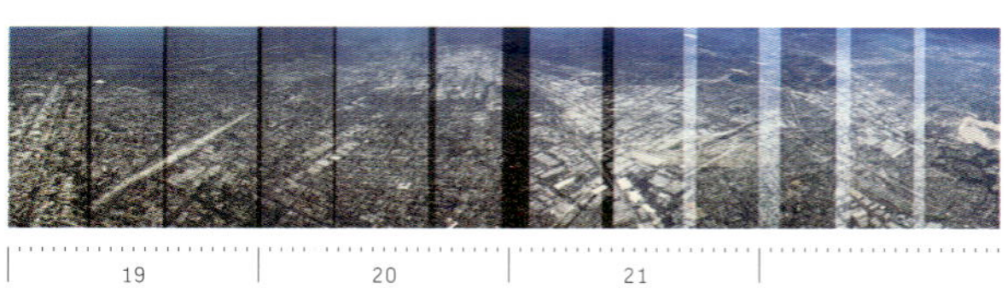

LA Land
19 x by 88 inches,
Edition of 10,
Lambda C-print, 2007

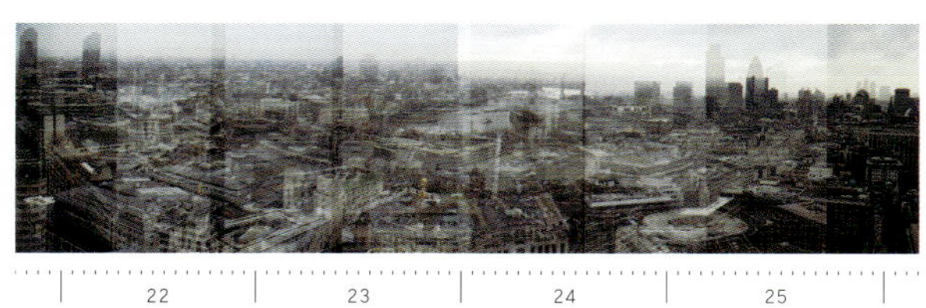

London
20 x 80 inches,
Edition of 10,
Lambda C-print, 2005

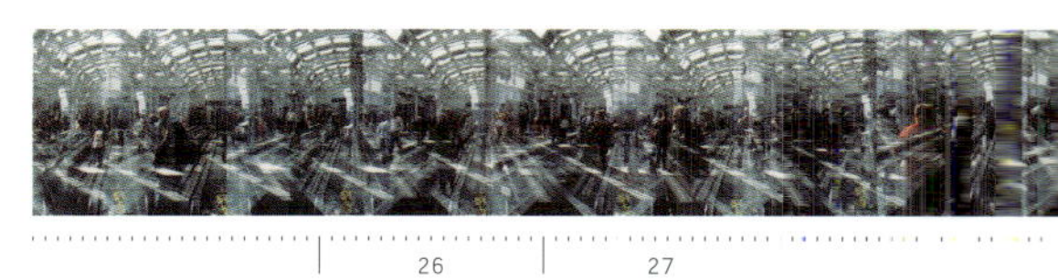

Converge
16 x 86 inches,
Edition of 10,
Lambda C-print, 2008

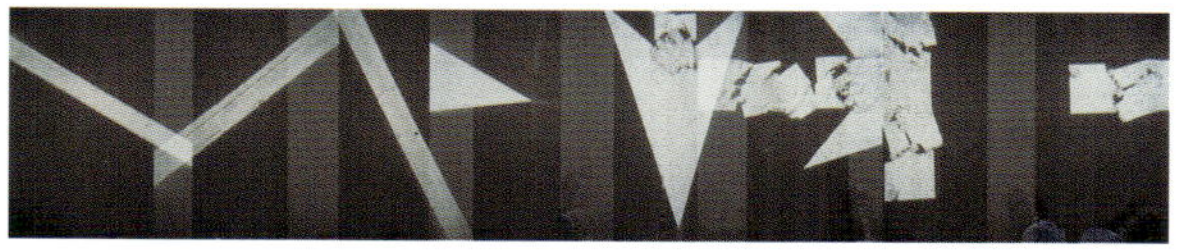

28 29 30 31

Signals
20 x 95 inches,
Edition of 10,
Lambda C-print, 2004

32 33

Exchange Street (New York City)
13 x 93 inches,
Edition of 10,
Lambda C-print, 2008

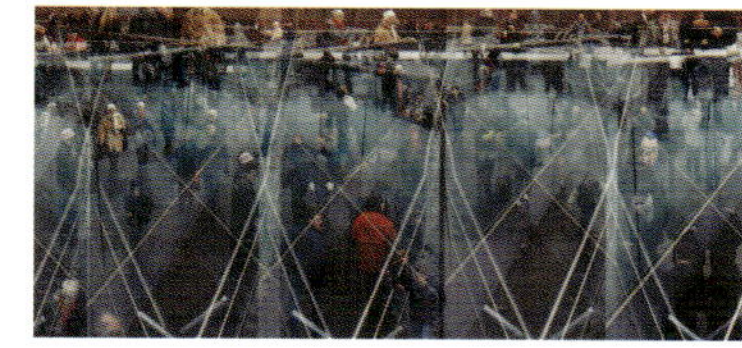

34 35

Foster Bridge (London)
20 x 48 inches,
Edition of 10,
Lambda C-print;2005

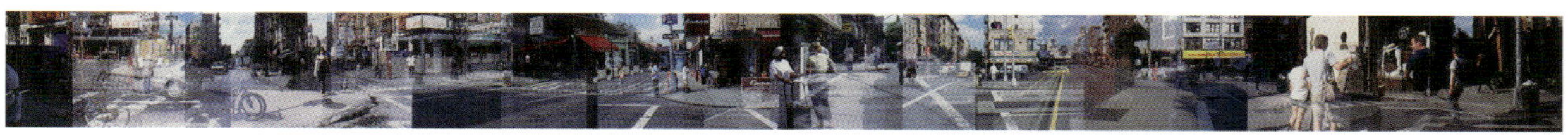

36 37 38 39 40 41

Lower East Side (NYC)
9 x 118 inches,
Edition of 10,
Lambda C-print, 2008

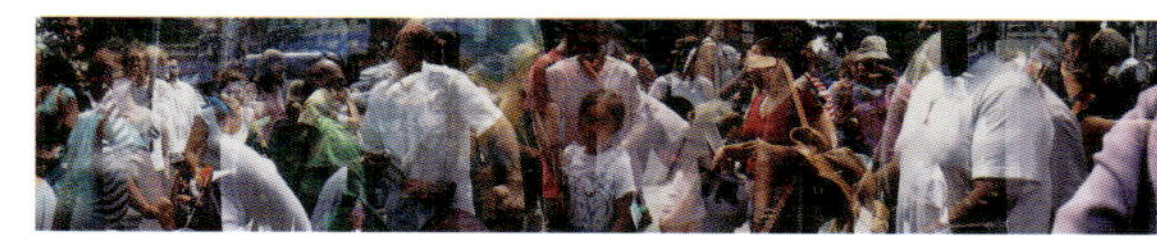

42 43

Brand Action (NYC)
20 x 78 inches,
Edition of 10,
Lambda C-print, 2008

46

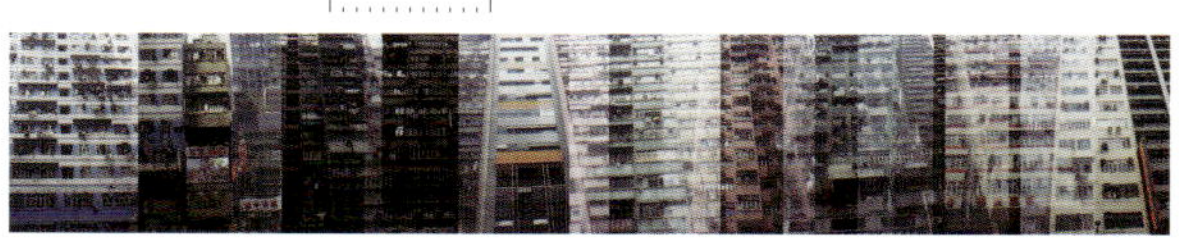

44 45

High Life (Hong Kong)
15 x 88 inches,
Edition of 10,
Lambda C-print, 2006

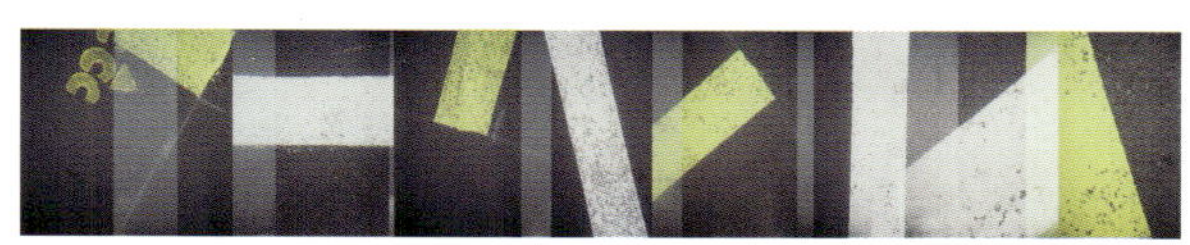

59 60 61

Kelly
16 x 84 inches,
Edition of 10,
Lambda C-print, 2004

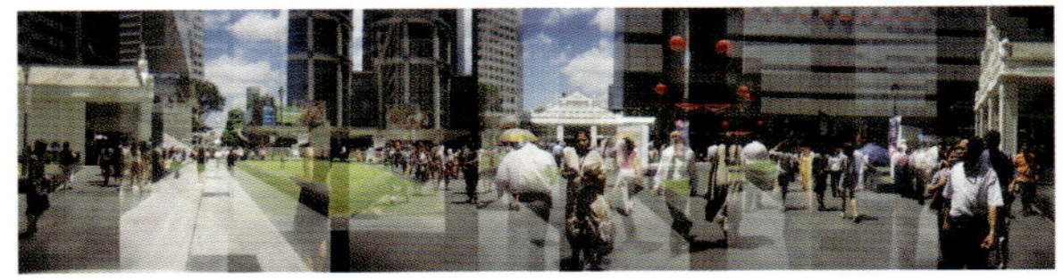

62 63

Singapore
21 x 80 inches,
Edition of 10,
Lambda C-print, 2005

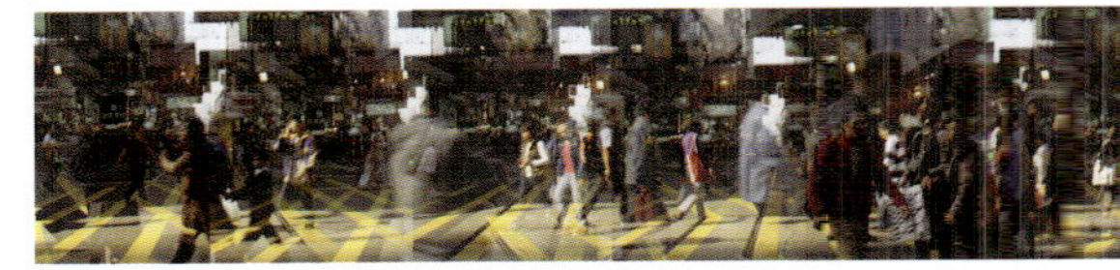

64 65

Causeway (Hong Kong)
20 x 88 inches,
Edition of 10,
Lambda C-print, 2006

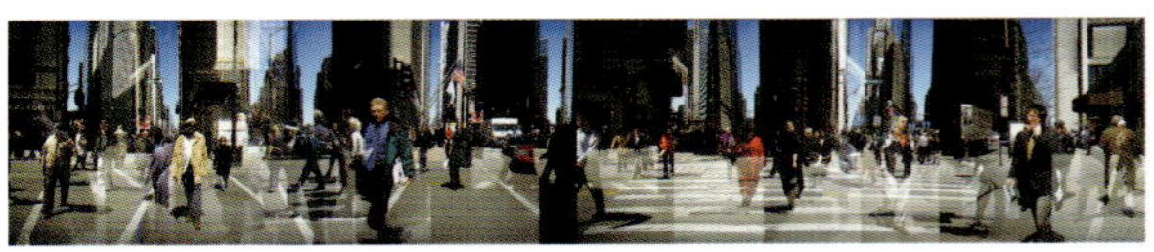

66 67

Chicago
19.5 x 96 inches,
Edition of 10,
Lambda C-print, 2007

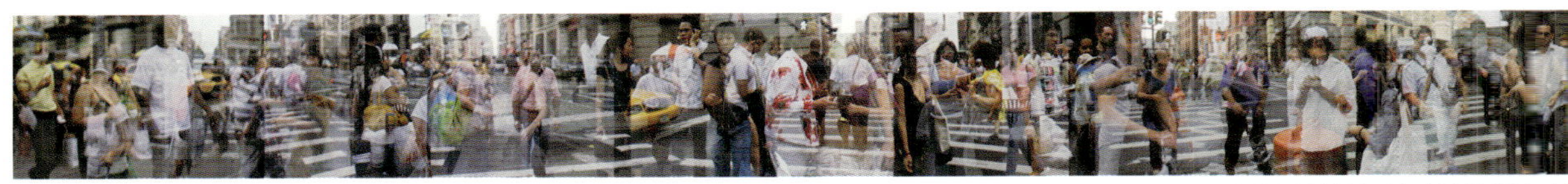

68 69 70 71 72

Broadway & Prince (New York City)
11.5 x 118 inches,
Edition of 10,
Lambda C-print, 2008

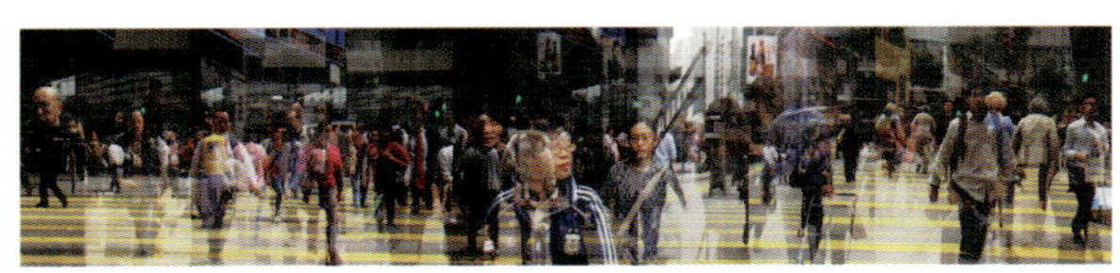

74 75

Central
20 x 90 inches,
Edition of 10,
Lambda C-print, 2006

78

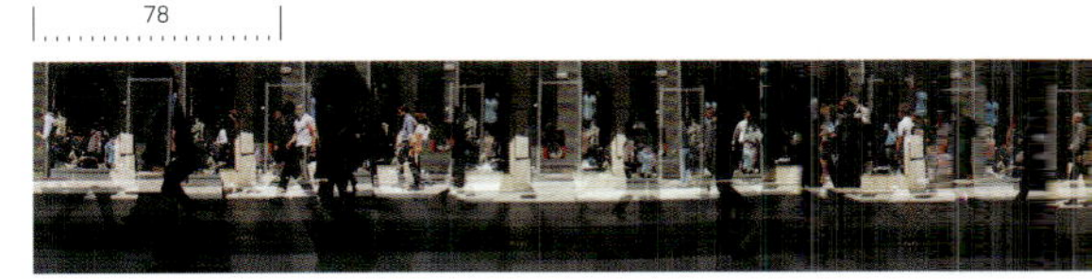

76 77

Pedway (Buenos Aires)
18 x 94 inches,
Edition of 10,
Lambda C-print, 2005

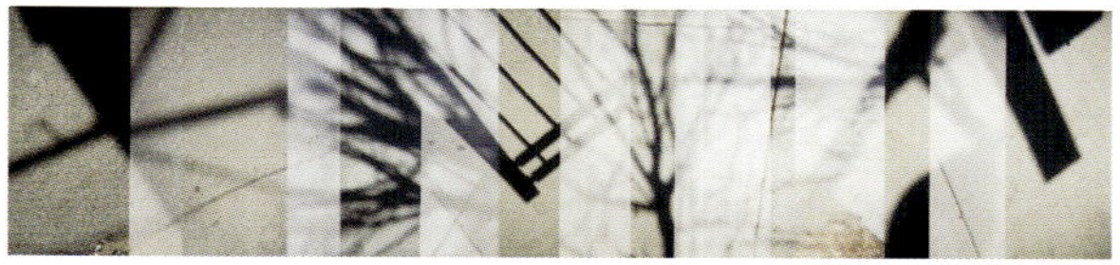

Shadows
20 x 90 inches,
Edition of 10,
Lambda C-print, 2004

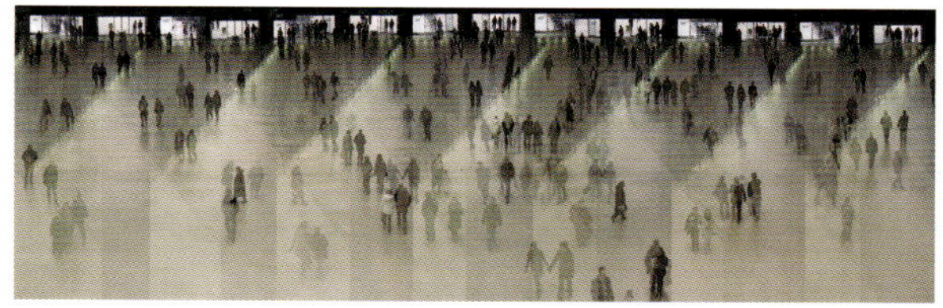

Tate Modern (London)
23 x 72 inches,
Edition of 10,
Lambda C-print, 2005

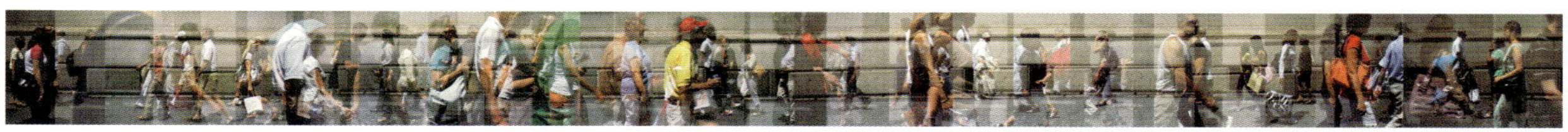

Shoppers (NYC)
9 x 118 inches,
Edition of 10,
Lambda C-print, 2008

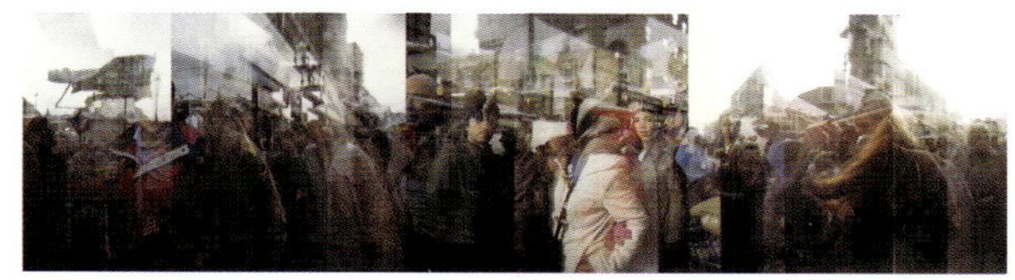

Portabello
23 x 88 inches,
Edition of 10,
Lambda C-print, 2005

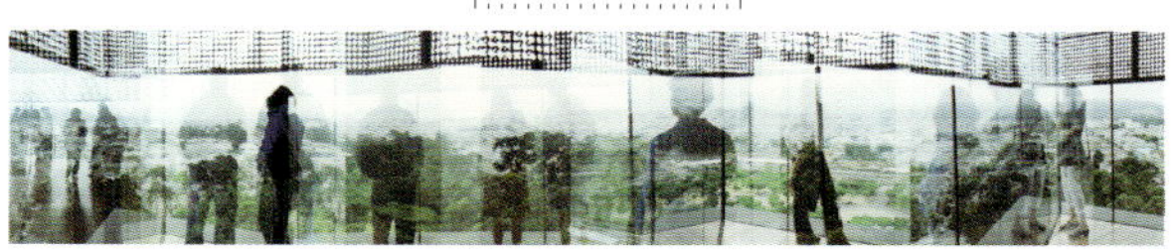

De Young
16 x 88 inches,
Edition of 10,
Lambda C-print, 2007

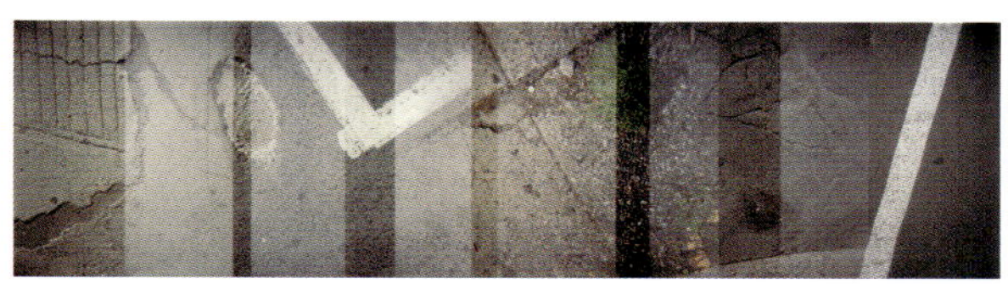

Chalk
20 x 84.5 inches,
Edition of 10,
Lambda C-print, 2004

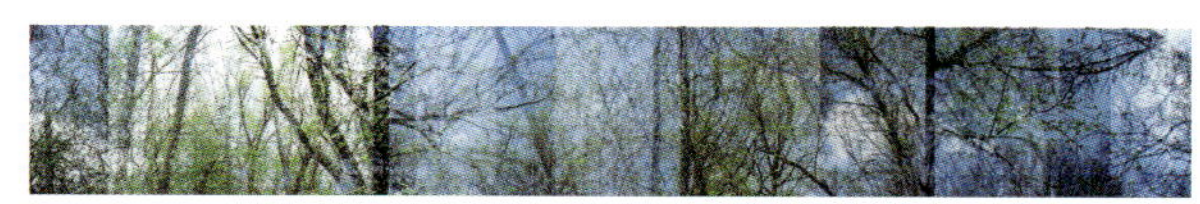

Native Spring
17 x 118 inches,
Edition of 10,
Lambda C-print, 2007

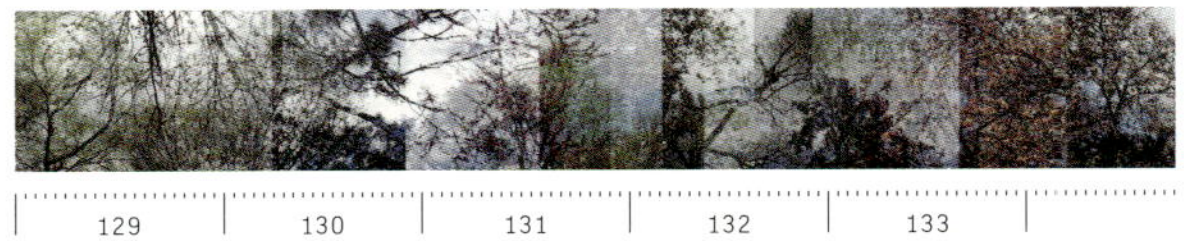

Spring (Illinois)
16 x 118 inches,
Edition of 10,
Lambda C-print, 2007

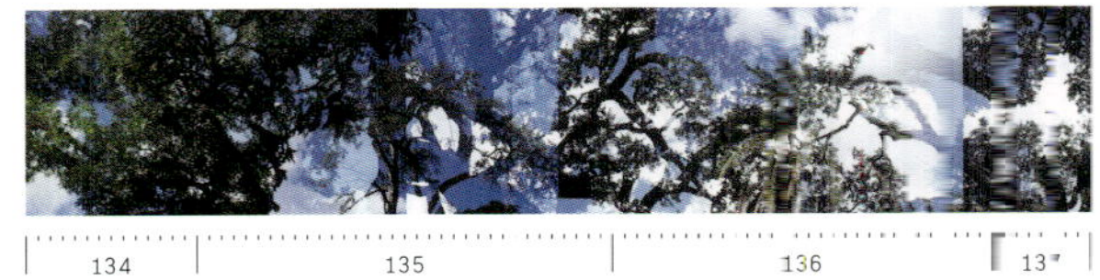

Threecosys
17 x 88 inches,
Edition of 10,
Lambda C-print, 2007

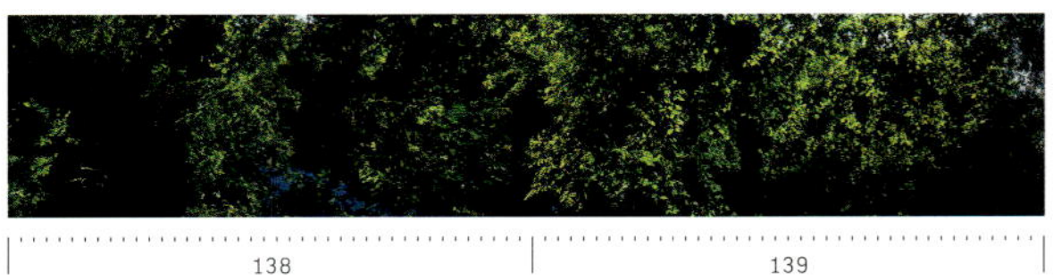

Bounty
20 x 82 inches,
Edition of 10,
Lambda C-print, 2006

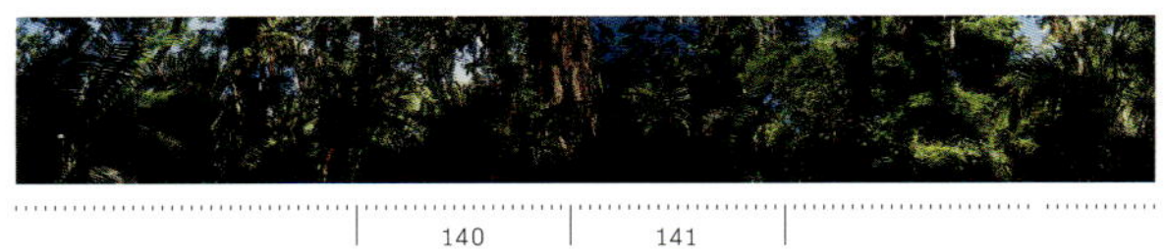

Sayulita (Mexico)
20 x 118 inches,
Edition of 10,
Lambda C-print, 2007

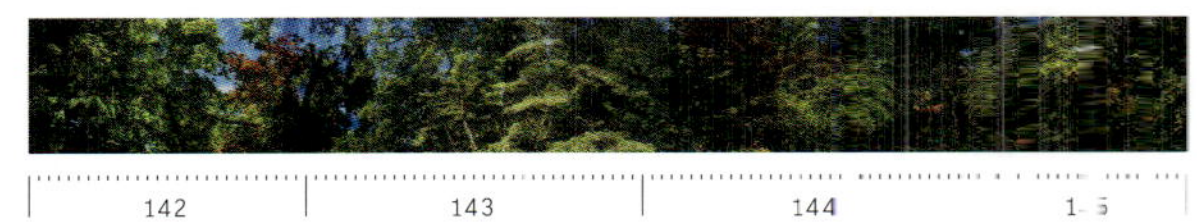

Butchart
14 x 118 inches,
Edition of 10,
Lambda C-print, 2007

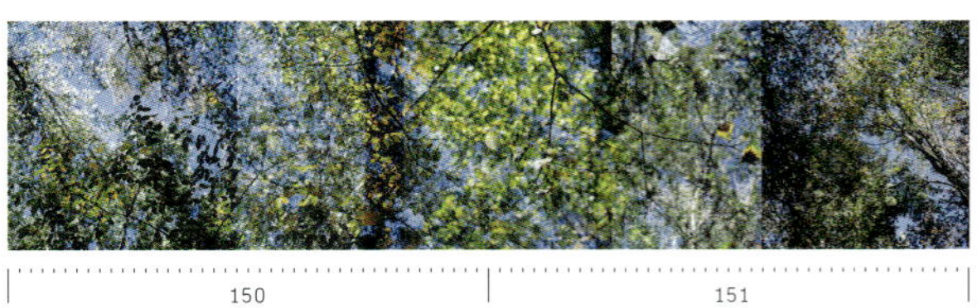

Flinch
20 x 85 inches,
Edition of 10,
Lambda C-print, 2007

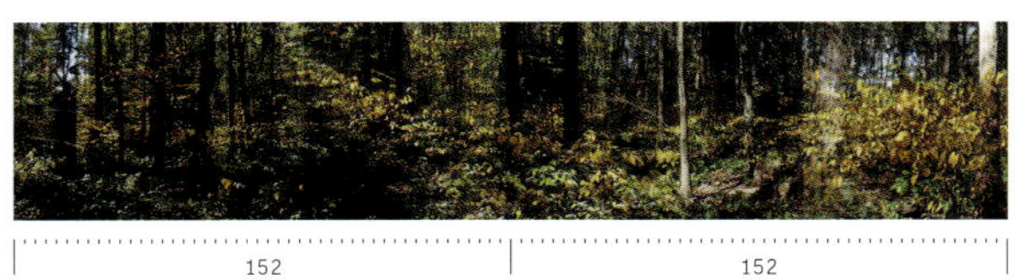

Forest Wetland (Indiana)
19 x 95 inches,
Edition of 10,
Lambda C-print, 2007

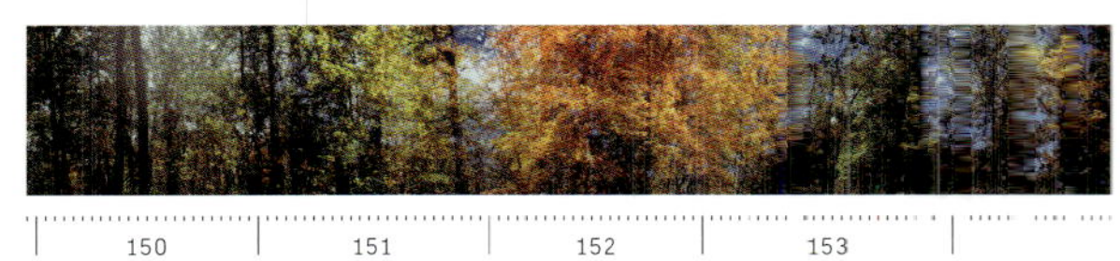

Maple Burst
18 x 118 inches,
Edition of 10,
Lambda C-print, 2008

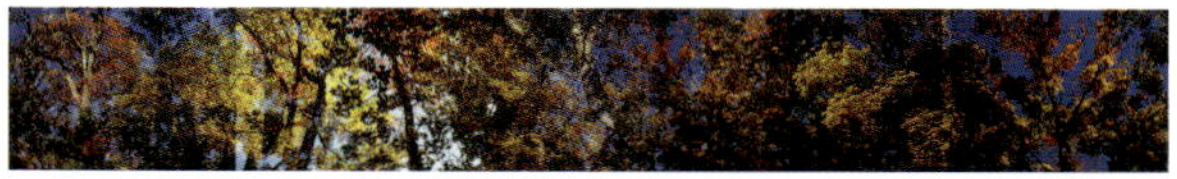

154 155

Fall 2008
16 x 118 inches,
Edition of 10,
Lambda C-print, 2008

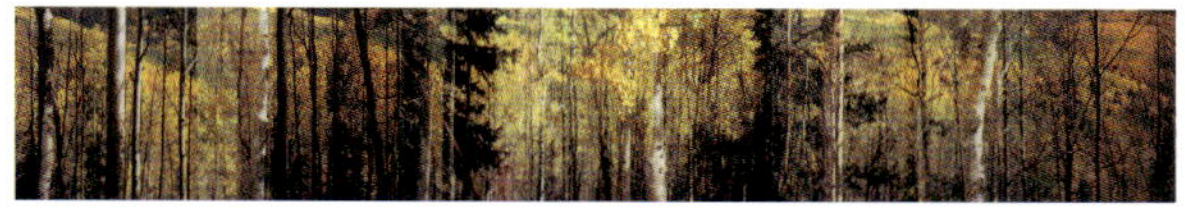

156 157

Aspens (New Mexico)
16 x 96 inches,
Edition of 10,
Lambda C-print, 2004

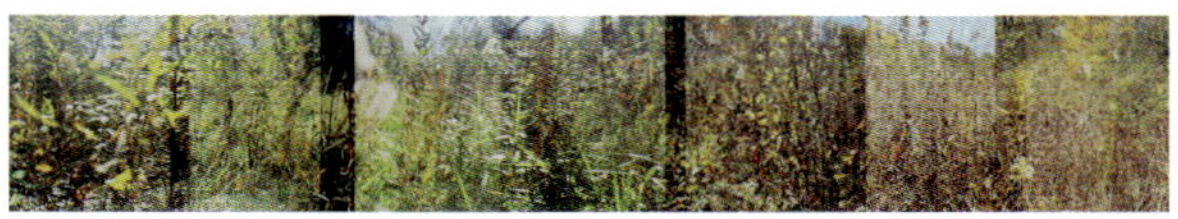

158 159

Artillery (Illinois)
16 x 96 inches,
Edition of 10,
Lambda C-print, 2007

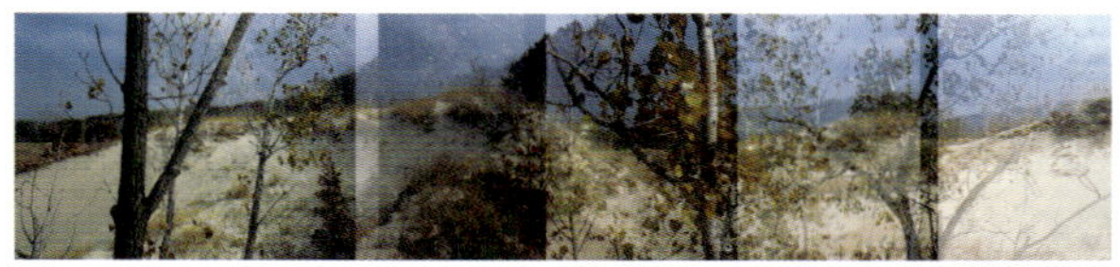

160 161

Dunes (Indiana)
20 x 89 inches,
Edition of 10,
Lambda C-print, 2007

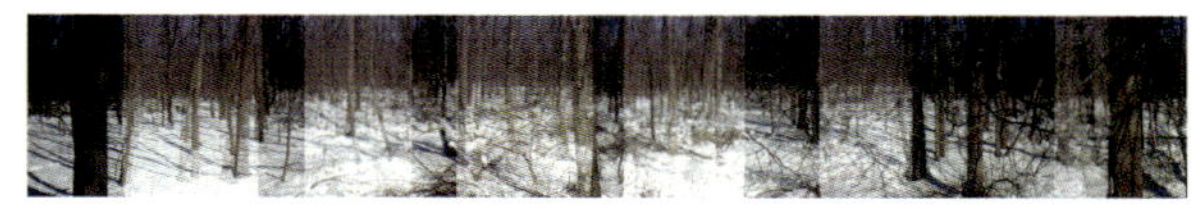

162 163

Gurnee
15 x 96 inches,
Edition of 10,
Lambda C-print, 2007

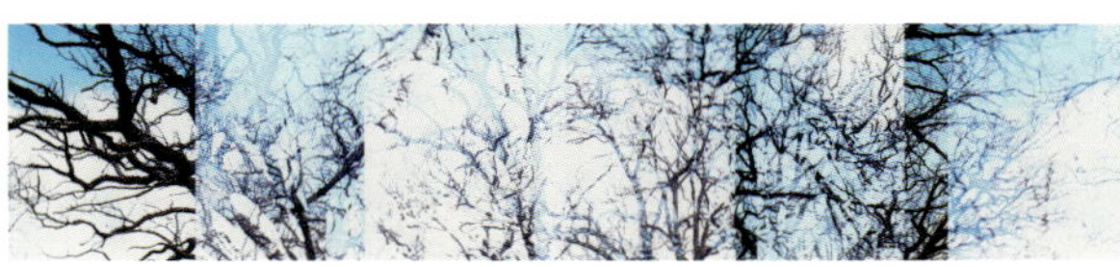

164 165

Winter Branches (Wisconsin)
20 x 96 inches,
Edition of 10,
Lambda C- print, 2006

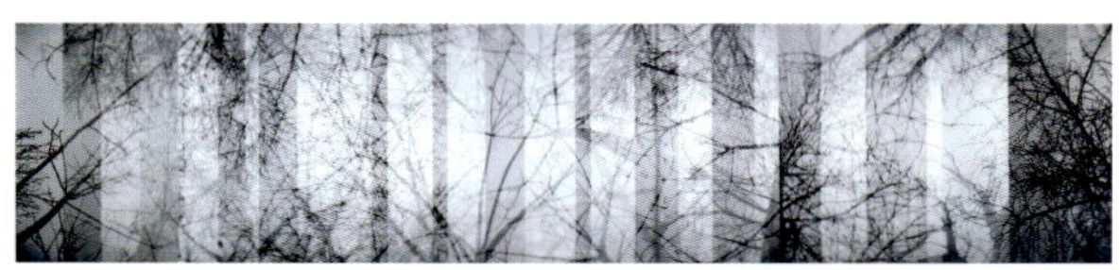

166 167

Neil's Branches
20 x 82 inches,
Edition of 10,
Lambda C-print, 2004

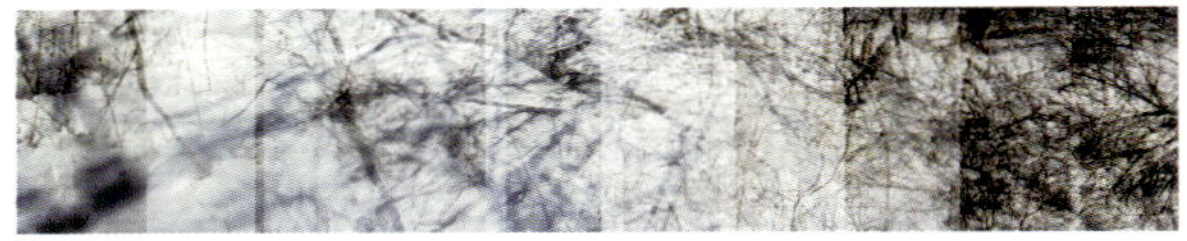

168 169

Snow Life
18 x 92 inches,
Edition of 10,
Lambda C-print, 2007

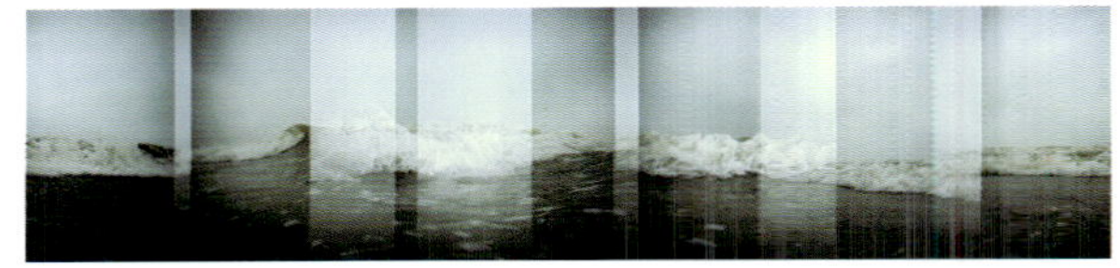

170 171

Mysto (Pacific Ocean)
22.5 x 88 inches,
Edition of 10,
Lambda C-print, 2006

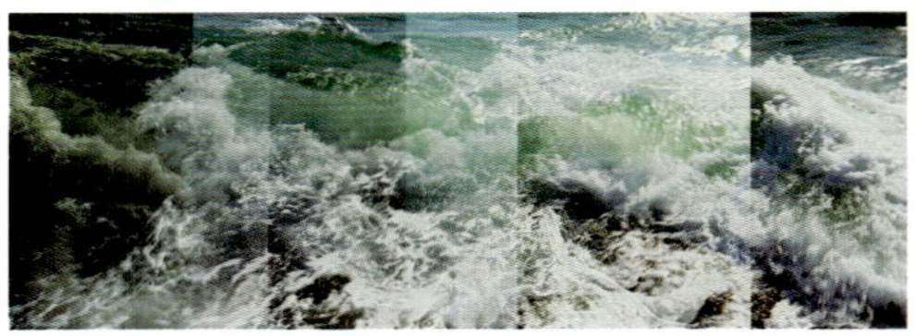

172 173

Break (Atlantic Ocean)
26.5 x 74 inches,
Edition of 10,
Lambda C-print, 2006

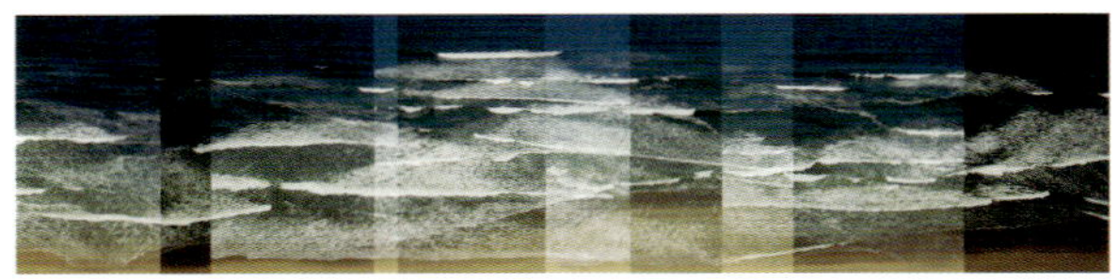

174 175

Longshore (Pacific Ocean)
20 x 86 inches,
Edition of 10,
Lambda C-print, 2006

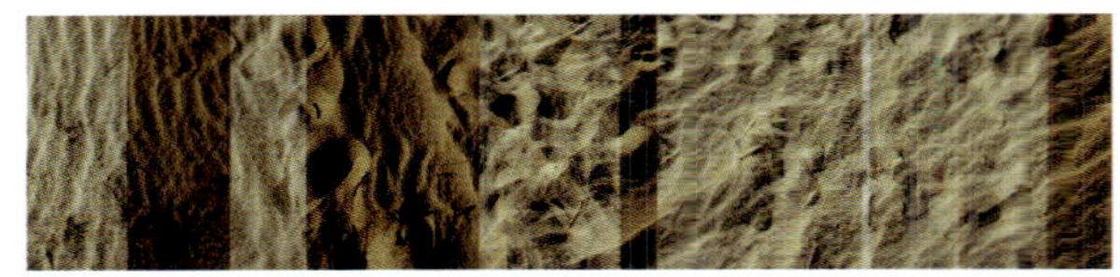

176 177

Desert (California)
17 x 72 inches,
Edition of 10,
Lambda C-print, 2007

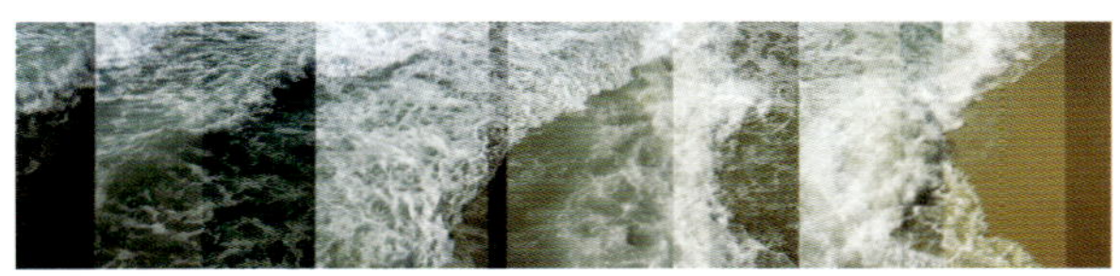

178 179

Backwash (Pacific Ocean)
19 x 90 inches,
Edition of 10,
Lambda C-print, 2006

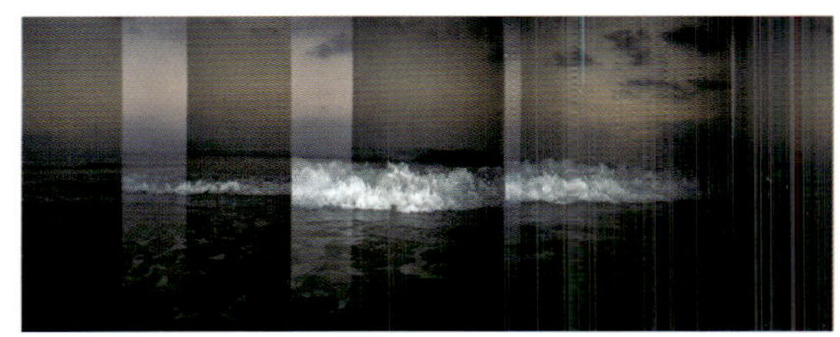

180 181

Cush (Pacific Ocean)
18.5 x 48 inches,
Edition of 10,
Lambda C-print, 2006

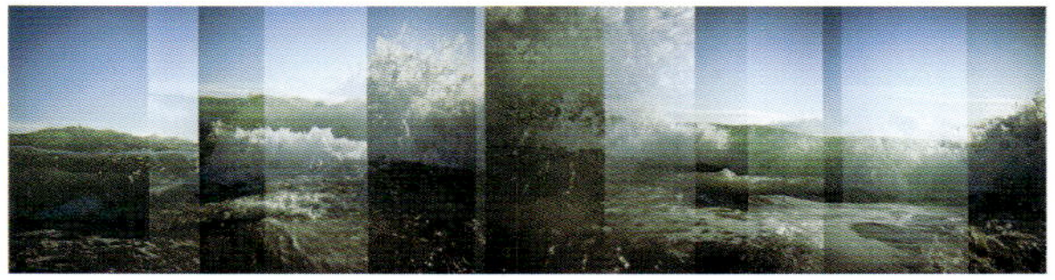

182 183

Axed (Pacific Ocean)
20 x 77 inches,
Edition of 10,
Lambda C-print, 2006

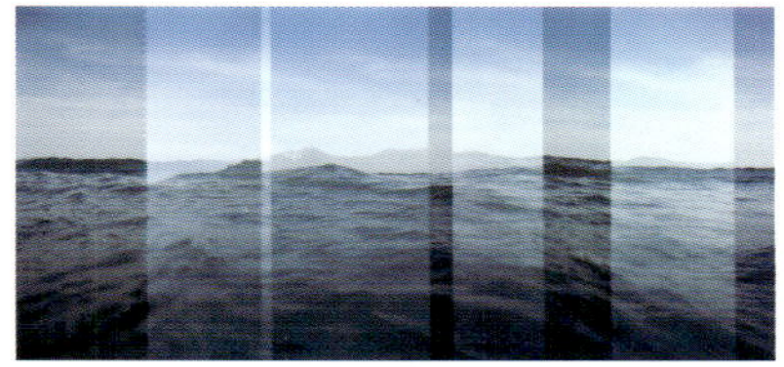

184 185

Line Up (Adriatic Sea)
24 x 52 inches,
Edition of 10,
Lambda C-print, 2006

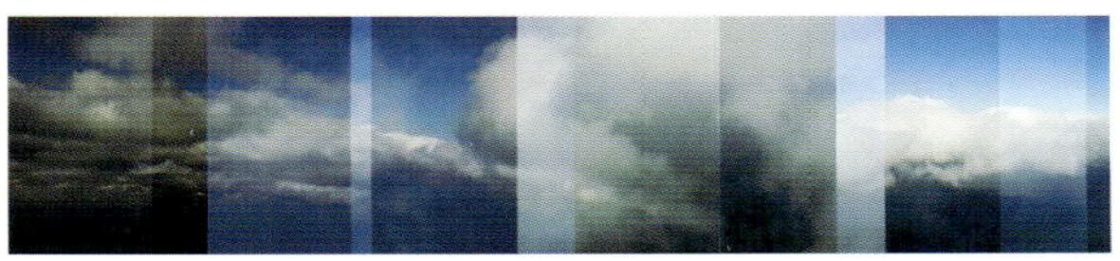

186 187

Homebound
18 x 84 inches,
Edition of 10,
Lambda C-print, 2006

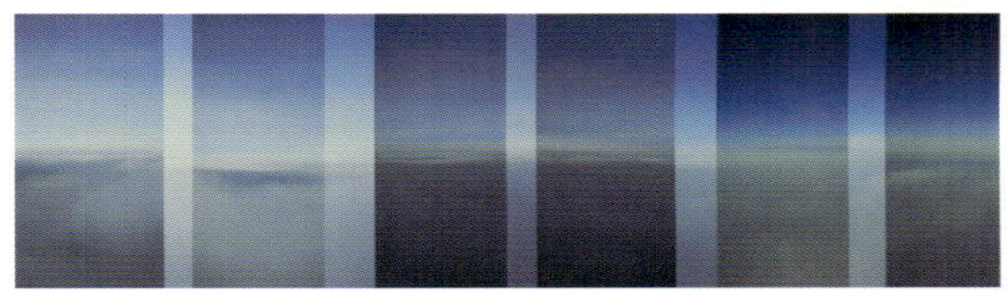

188 189

Aria
25 x 90 inches,
Edition of 10,
Lambda C-print, 2006

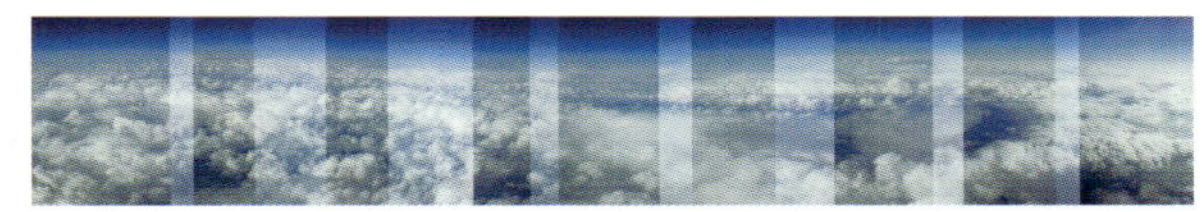

190 191 192 193 194

System
15.5 x 96 inches,
Edition of 10,
Lambda C-print, 2006

195 196 197 198

Sunburst
19 x 94 inches,
Edition of 10,
Lambda C-print, 2007

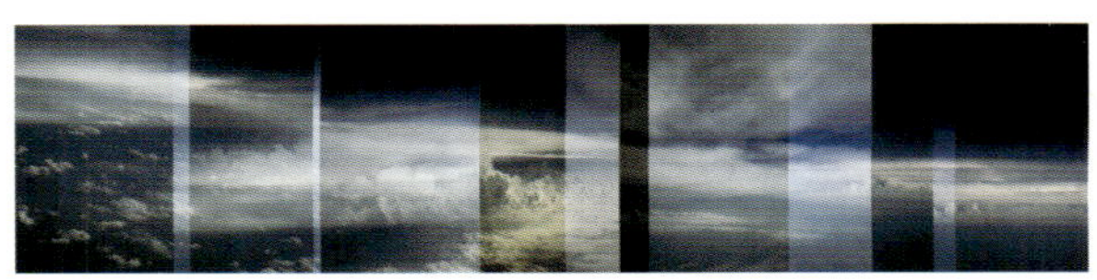

199 200 201

Etheria
22 x 96 inches,
Edition of 10,
Lambda C-print, 2007

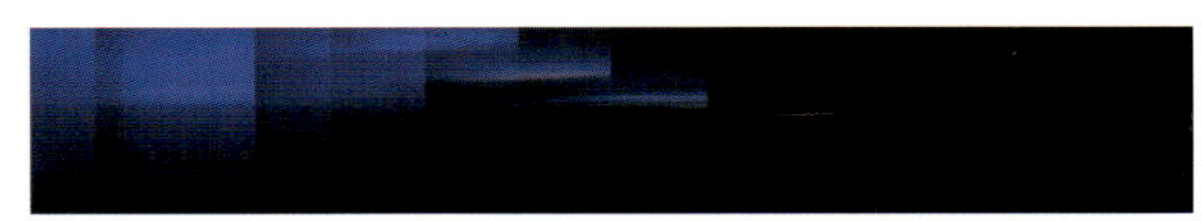

202 203 204 205

Nightfall
17.5 x 94 inches,
Edition of 10,
Lambda C-print, 2007

For Kai Marlow Fogelson

Acknowledgments

Heartfelt thanks to Eiren Caffall, Derrick Jensen and Bridgette R. McCullough Alexander for their thought-provoking essays. Thanks to Tim Hartford and Ron Alikpala for tireless and sensitive design. Many thanks also to Caitlin Bauler, Dan De Los Monteros, Christine DiThomas, David Castillo, Shashi Caudill, Maya Piergies, Mike Riley, Kristi McGuire, University of Chicago Press, Irene Fogelson, and the entire family.

Design: Hartford Design
Production: Caitlin Bauler
Editing: Christine DiThomas

dougfogelson.com
front40press.com

Fogelson, Doug.

The time after / Doug Fogelson ; [essays by Derrick Jensen, Eiren Caffall, and Bridgette R. McCullough]. – Chicago : Front Forty Press, 2008.
p. ; cm.

ISBN: 978-0-9778689-5-7
Includes index.

1. Environment (Art) 2. Human beings–Effect of environment on. 3. Nature–Effect of human beings on. 4. Social ecology. 5. Human ecology. 6. Images, Photographic. 7. Photography, Artistic. I. Jensen, Derrick II. Caffall, Eiren. III. McCullough Alexander, Bridgette R. IV.

N6494.E6 F64 2008
709.04-dc22 2008

ISBN 0-9778689-9-0

First Edition 2009